FRONTIERS IN STEM CELL RESEARCH

FRONTIERS IN STEM CELL RESEARCH

JULIA M. SPANNING
EDITOR

Nova Science Publishers, Inc.
New York

For permission to use material from this book please contact us:
Telephone 631-231-7269; Fax 631-231-8175
Web Site: http://www.novapublishers.com

LIBRARY OF CONGRESS CATALOGING-IN-PUBLICATION DATA

Frontiers in stem cell research / Julia M. Spanning (editor).
p. cm.
Includes index.
ISBN 13 978-1-60021-294-9
ISBN 10 1-60021-294-8
1. Stem cells--Research. 2. Embryonic stem cells--Research. I. Spanning, Julia M.
QH588.S83F76 2006
616'.02774--dc22 2006018904

Published by Nova Science Publishers, Inc. ✢ New York

CONTENTS

PREFACE

Among the many applications of stem cell research are nervous system diseases, diabetes, heart disease, autoimmune diseases as well as Parkinson's disease, end-stage kidney disease, liver failure, cancer, spinal cord injury, multiple sclerosis, Parkinson's disease, and Alzheimer's disease. Stem cells are self-renewing, unspecialized cells that can give rise to multiple types all of specialized cells of the body. Stem cell research also involves complex ethical and legal considerations since they involve adult, fetal tissue and embryonic sources. This new book brings together leading research from throughout the world in this frontier field.

At least two areas of the adult mammalian brain continue to produce new neurons throughout the adult lifespan: the hippocampus and the olfactory bulb. These new neurons are derived from stem cells located in the subgranular zone of the hippocampus, and the anterior portion subventricular zone surrounding the lateral ventricles, respectively. In the hippocampus, precursor cells divide in the subgranular zone and migrate relatively short distances into the overlying granule cell layer, where they differentiate into new granule cell neurons. In contrast, neuronal precursors destined for the olfactory bulb must migrate long distances first tangentially through the rostral migratory stream, and then radially into the outer layers of the olfactory bulb where they differentiate into two types of inhibitory interneurons: granule cells, and periglomerular cells. This migration requires an ongoing rearrangement of the cytoskeleton as well as integration of chemo-attractive and repulsive cues from the environment to result in directed movement of the cell along the appropriate pathways. Molecules proposed to play a role in the migration of neuronal precursors along the rostral migratory stream include polysialated neural cell adhesion molecule, netrins, semaphorins, slits, ephrins, and neuregulins. In addition, two molecules in the extracellular matrix, reelin and Tenascin-R, as well as the secreted factor prokineticin-2 have been proposed to key roles in the transition from radial to tangential migration. In chapter 1, the authors discuss what is known about the factors regulating the migration of adult neuronal precursors to the olfactory bulb, as well as recent unpublished data from their laboratory suggesting possible roles of members of the ERM (ezrin-radixin-moesin), cadherin, and catenin families in this process.

Adult bone marrow stem cell therapy has been the subject of tremendous research in recent years. Adult marrow stem cells (MSCs) participate in the regeneration of the myocardium after acute ischemic injuries. This process, in part, in part, involves migration of MSCs from remote regions to sites of injuries, as well as milieu-dependent differentiation initiated through a complex interaction of signals. This process of differentiation ultimately lead to an improvement of cardiac function. Therapeutically, various approaches to the delivery of these MSCs have been proposed. However, in order to develop the most effective delivery regimen, insights into the homing-mechanisms are crucial. These mechanisms involve a complex interaction between cellular signaling and paracrine, or endocrine factors.

In chapter 2, the authors found that acute ischemic insult provided a strong signaling mechanism that directs homing of MSCs residing in bone marrow after systemic transfusion of MSCs. This signaling mechanism was reproduced in a subsequent study to investigate the presence or absence of homing signals in a chronic myocardial injury model. Interestingly, in subjects with chronic ischemic injury, the homing of MSCs was absent. In a subsequent study, the authors investigated whether these homing signals in the recruitment of MSCs for myocardial regeneration were present in immature hearts, and whether normal physiologic growth provided similar homing signals in the recruitment of MSCs. They found that homing signals were present even in the immature myocardium after acute injury to effect the identical regeneration process in their adult counterpart. However, normal physiologic growth provided no recruiting signals to MSCs. Insights into the cellular and molecular basis of homing mechanisms have been shed in recent years. Implicated in the process of MSC recruitment to the injured myocardium, myocardial regeneration and angiogenesis are G-CSF, M-CSF, VEGF, SCF-1, among many others. Although the exact interplay between various homing signals remains to be defined, the results of their studies have important implications on stem cell therapeutics in myocardial regeneration. Further studies to elucidate the homing mechanisms in acute myocardial injury will be required.

Embryonic stem cells (ESCs) are undifferentiated cells, which are typically derived from the inner cell mass of a blastocyst-stage embryo. In culture, these cells have the capacity to self-renew in an undifferentiated state but also may differentiate into cell types representing the three embryonic germ lineages, thus revealing their pluripotent potential. These cells could be of value for creating in-vitro culture systems and animal models that may be used to study human genetic disease, gene function, drug discovery, developmental biology, and differentiation events. Chapter 3 summarizes different techniques for the derivation and maintenance of undifferentiated ESCs, in particular with emphasis on human ESCs. In the second part, issues concerning self-renewal and pluripotency are explored. In addition, accumulated data regarding the in vitro differentiation potential of ESCs is evaluated. The last part of this review outlines basic criteria, which may result in the use of ESCs in future clinical applications.

Chapter 4 examines two public policy decisions and their subsequent effect upon health care. The first decision was the proposed Lethal Drug Abuse prevention Act of 1998 otherwise known as the Hyde-Nichols bill. While the bill ultimately died in the Congress the debate it generated has allowed scholars and policymakers alike to assess its impact upon health care, specifically the healthcare associated with pain management. The bill had political and ideological motivations in that its main purpose was to nullify the state of Oregon's physician assisted suicide legislation. The subsequent debate illustrates the problem of legislating from the perspective of political ideology and not pure science. The second decision is that of President Bush regarding the usage of stem cells in basic biological and medical research. The decision to use existing stem cells has already been surpassed by the needs of scientific research. This has touched off a vigorous debate concerning the need for an expanded policy regarding the use of stem cells. Both decisions pose problems for basic research and health care. The authors note these potential difficulties and draw conclusions concerning the role of political ideology in the health care field. They also present suggestions for alleviating some of these difficulties.

Skin stem cell biology has been a focus of increasing interest in life science. Given its easy accessibility, the broad developmental potential of stem cell populations from skin has provided more wide practical implications. Skin is the biggest organ in the body and is constantly renewed. This process of renewal necessitates the presence of adult stem cells. To date, research has focused on the application of epidermal stem cells. Recent research also shows there are at least three different types of other stem cells (mesenchymal, neural and hematopoietic stem cells) that have been identified from the dermis and hair follicles. Mesenchymal stem cells from bone marrow are the best studied adult multipotent stem cells and have shown promising therapeutic implications in regenerative medicine. The multipotent stem cells from the dermis have the potential to produce distinct cell lineages, such as fibroblasts, follicle dermal cells, endothelial cells, marrow stromal cells, osteocytes, chondrocytes, muscle cells, and even neural cells. Recently research indicates that the dermis may represent a novel source of mesenchym stem cells. In chapter 5, recent findings regarding the biological properties and potential clinical use of dermis-derived multipotent stem cells with mesenchymal potential are summarized. Recent advances of the neural stem cells and hematopoietic stem cells from the dermis are also discussed.

Type 1 diabetes mellitus (T1DM) is a disorder of a single cell type, the insulin-secreting pancreatic β-cell, and as such is particularly amenable to treatment by cell replacement therapy. For this reason, T1DM has received much attention recently as a potential target for the emerging science of stem cell medicine. In this autoimmune disease, the pancreatic β-cells are selectively and irreversibly destroyed by autoimmune assault. Advances in islet transplantation procedures now mean that individuals with T1DM can be cured by human islet transplantation. A major drawback in this therapy is the limited availability of donor islets, and the search for substitute transplant tissues has intensified in the last few years. This review in chapter 6, will summarise recent progress in using stem cell populations for generating substitute β-cells for transplantation therapy. The relative merits of stem cells from different sources will be considered, and the major technical obstacles to the production of reliable and reproducible stem cell to β-cell differentiation protocols will be addressed.

Since the first experiments of cell transplantation into heart were performed in the early 1990s, the identification of adult stem cells has triggered attempts to regenerate damaged heart tissue by cellular transplantation. Until recently, a multitude of adult stem or progenitor cells from various tissues have been proposed to meet this end. Bone marrow, especially, has emerged as the most promising source for stem and progenitor cells because bone marrow, besides the organ of hematopoietic maintenance, contains a complex assortment of stem and progenitor cells. A large body of provocative experimental evidence for vascular and myocardial regeneration by these cells has furthered enthusiasm for their use, yet many questions remain unanswered in this new field of research regarding the therapeutic potential and the mechanisms responsible for the observed therapeutic effects. In this review article in chapter 7, the authors will discuss the therapeutic capacity of currently available representative bone marrow-derived stem and progenitor cells for treating ischemic heart diseases.

In: Frontiers in Stem Cell Research
Editor: Julia M. Spanning, pp. 1-38

ISBN 978-1-60021-294-8

Chapter I

MEMBRANE-CYTOSKELETAL INTERACTIONS IN THE REGULATION OF NEURONAL MIGRATION IN THE ADULT BRAIN

Robert D. Beech*

Yale University, Department of Psychiatry, 300 George St., New Haven, CT 06511, USA

ABSTRACT

At least two areas of the adult mammalian brain continue to produce new neurons throughout the adult lifespan: the hippocampus and the olfactory bulb. These new neurons are derived from stem cells located in the subgranular zone of the hippocampus, and the anterior portion subventricular zone surrounding the lateral ventricles, respectively. In the hippocampus, precursor cells divide in the subgranular zone and migrate relatively short distances into the overlying granule cell layer, where they differentiate into new granule cell neurons. In contrast, neuronal precursors destined for the olfactory bulb must migrate long distances first tangentially through the rostral migratory stream, and then radially into the outer layers of the olfactory bulb where they differentiate into two types of inhibitory interneurons: granule cells, and periglomerular cells. This migration requires an ongoing rearrangement of the cytoskeleton as well as integration of chemo-attractive and repulsive cues from the environment to result in directed movement of the cell along the appropriate pathways. Molecules proposed to play a role in the migration of neuronal precursors along the rostral migratory stream include polysialated neural cell adhesion molecule, netrins, semaphorins, slits, ephrins, and neuregulins. In addition, two molecules in the extracellular matrix, reelin and Tenascin-R, as well as the secreted factor prokineticin-2 have been proposed to key roles in the transition from radial to tangential migration. In this review, I discuss what is known about the factors regulating the migration of adult neuronal precursors to the olfactory bulb, as well as recent unpublished data from our laboratory suggesting

* Correspondence concerning this article should be addressed to Robert D. Beech, MD, Ph.D. Assistant Professor Yale University, Department of Psychiatry, 300 George St., 8304C, New Haven, CT 06511.

possible roles of members of the ERM (ezrin-radixin-moesin), cadherin, and catenin families in this process.

Keywords: cell migration, subventricular zone, rostral migratory stream, polysialated neural cell adhesion molecule, netrin, semaphorin, slits, ephrin, neuregulin, extracellular matrix, reelin, Tenascin-R, prokineticin-2, ERM (ezrin-radixin-moesin), cadherin, and catenin.

INTRODUCTION

While many areas of the adult CNS contain stem cells that are capable of giving rise to neurons *in vitro,* only stem cells located in the hippocampal dentate gyrus and the subventricular zone appear to give rise to new adult neurons *in vivo*. This is thought to be the result of bi-directional communication between multipotent stem cells and the three-dimensional local milieu or niche that they inhabit (Cheng et al., 2005b; Spradling et al., 2001; Wurmser et al., 2004). The concept of a unique niche that serves to maintain the "stemness" of progenitor cells was first proposed for hematopoetic progenitors in the late 1960's (Wolf and Trentin, 1968), and more recently has been applied to progenitors of adult generated neurons (Palmer et al., 2000). In vitro experiments have indicated important roles for both glial (Lim and Alvarez-Buylla, 1999; Song et al., 2002) and endothelial (Shen et al., 2004) cells in this process. However, the precise molecular components and signaling pathways that are required to produce and maintain this microenvironment remain incompletely characterized.

In the hippocampus, precursor cells divide in the subgranular zone and migrate relatively short distances into the overlying granule cell layer, where they differentiate into new granule cell neurons (Cameron and Gould, 1994; Eriksson et al., 1998; Kornack and Rakic, 1999; Kuhn et al., 1996). In contrast, adult neuronal precursors destined for the olfactory bulb must migrate long distances first tangentially through the rostral migratory stream (RMS), and then radially into the outer layers of the olfactory bulb where they differentiate into two types of inhibitory interneurons: granule cells, and periglomerular cells (figure 1) (Lois and Alvarez-Buylla, 1994; Luskin, 1993). Adult generated neurons in the olfactory bulb appear to have distinct physiological characteristics, including increased sensitivity to activity-dependent regulation of surivival, compared to those generated by the early post-natal RMS (Corotto et al., 1994; Giachino et al., 2005; Lemasson et al., 2005; Winner et al., 2002). Work in our laboratory suggests that while precursors for both granule cells and periglomerular cells are located in the SVZ, there may separate pools of stem cells which give rise to each of these two neuronal populations (Beech et al., 2004). The transcription factor Pax6 appears to play a key role in specifying the facte of prospective dopamingeric periglomerular neurons (Hack et al., 2005; Kohwi et al., 2005). The mechanisms determining the fate of other types of newly generated neurons have not yet been determined.

Tangential migration in the adult RMS differs from radial migration of cortical neuronal precursors, which occurs during embryonic development, in that it does not require radial glial or axonal guides (Lois et al., 1996; Marin and Rubenstein, 2001; Marin and Rubenstein, 2003). However, migrating cells in the SVZ and RMS are ensheathed by a network of astrocytic cells (glial tubes) (Peretto et al., 1997). The cells comprising the glial tubes serve both as the precursors of, and a physical channel for, the migrating cells (Alvarez-Buylla and Garcia-Verdugo, 2002; Doetsch et al., 1999). Migration of neuronal precurosors through the RMS requires an ongoing rearrangement of the cytoskeleton (Ponti et al., 2004) as well as integration of chemo-attractive and repulsive cues from the environment to result in directed movement of the cell along the appropriate pathways.

Several recent papers have addressed the regulation of stem-cell proliferation in the SVZ by external factors (Hack et al., 2005; Liu et al., 2005; Lledo and Saghatelyan, 2005; Matarredona et al., 2005; Palma et al., 2005; Tonchev et al., 2005; Wang et al., 2005). The reader is referred to these papers for a more detailed discussion of factors involved in the regulation of proliferation in the SVZ. In this review I discuss what is known about the factors regulating the migration of adult neuronal precursors to the olfactory bulb, as well as recent unpublished data from our laboratory suggesting possible roles of specific membrane-cytoskeletal cross-linking proteins including members of the ERM (ezrin-radixin-moesin), cadherin, and catenin families in this process.

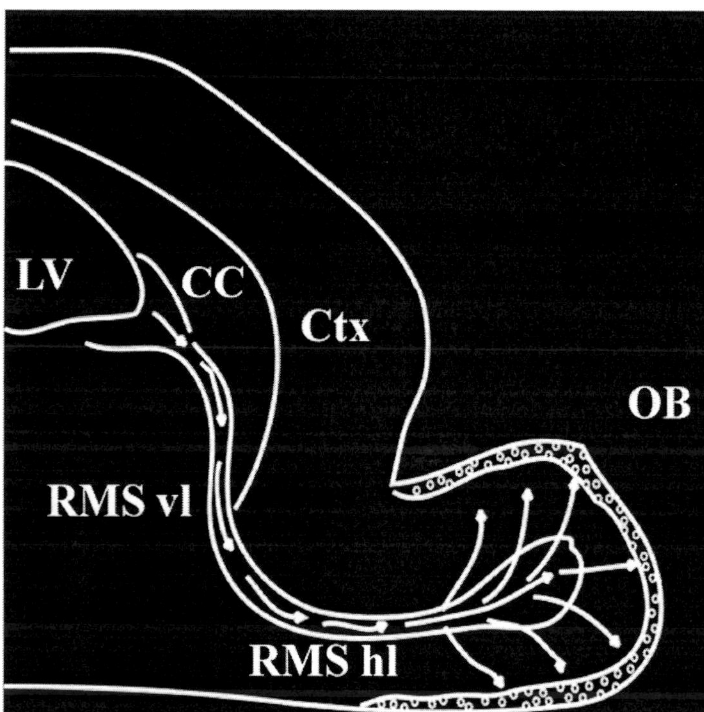

Figure 1: Schematic of the mouse forebrain illustrating the migratory pathway of SVZa derived neuroblasts en route to the olfactory bulb. Neuronal precursors destined for the olfactory bulb must migrate long distances first tangentially through the rostral migratory stream, and then radially into the outer layers of the olfactory bulb where they differentiate into two types of inhibitory interneurons: granule cells, and periglomerular cells . LV-Lateral Ventricle, CC- Corpus Callosum, CTx - Cerebral Cortex, RMS vl - Vertical Limb of Rostral Migratory Stream, RMS hl - Horizontal Limb of Rostral Migratory Stream, OB - Olfactory Bulb (adapted from (Coskun and Luskin, 2002)).

Basic Mechanisms of Cell Migration

Cell migration can be thought of as a cyclical process (Lauffenburger and Horwitz, 1996; Sheetz et al., 1999). Essential steps in this process include: (1) extension of the leading edge; (2) adhesion to matrix contacts; (3) contraction of the cytoplasm; (4) release from contact sites and recycling of membrane receptors from the rear to the front of the cell (figure 2). Each of these phases, in turn, is dependent upon multiple cyclical biochemical processes, many of which are regulated by members of the Rho family of GTPases (Raftopoulou and Hall, 2004; Ridley, 2001). Some of these pathways are described in more detail below.

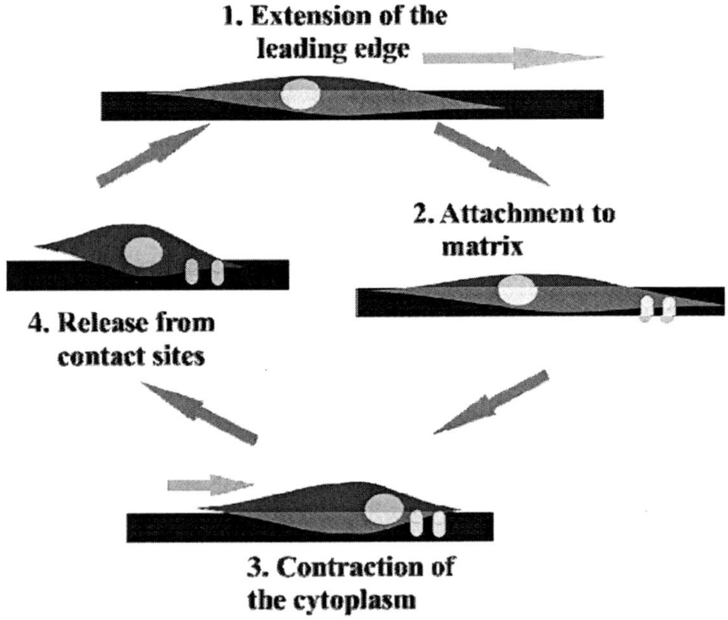

Figure 2: Cell migration as a cyclical process. Essential steps in the process of cell migration include: (1) extension of the leading edge; (2) adhesion to matrix contacts; (3) contraction of the cytoplasm; (4) release from contact sites and recycling of membrane receptors from the rear to the front of the cell. Each of these phases, in turn, is dependent upon multiple cyclical biochemical processes (see text) (adapted from (Sheetz et al., 1999)).

1. Extension of the Leading Edge is Regulated by Rho Family Members Rac1 and Cdc 42

Extension of the leading edge is driven by polymerization of actin filaments and bundling of these filaments into networks (Lodish et al., 2003) (figure 3). The growth of actin filaments occurs by the addition of ATP-actin subunits at the barbed, or (+) end of actin filaments, and the loss of subunits at the pointed, or (-) end. This process has been described as "actin treadmilling" (Pollard et al., 2000; Pollard and Borisy, 2003). Branching of filaments is mediated by activation of the Arp 2/3 complex, which binds to the sides or tip of existing filaments and induces the formation of a new daughter filament at a 70^0 angle from the mother filament (Welch and Mullins, 2002). These branches are stabilized by cortactin, and stabilization of the entire network occurs through cross-linking of the individual filaments by filamen A and α-actinin. Growth of fibers is terminated by the action of capping proteins, while members of the ADF/Cofilin family of proteins regenerate the pool of free actin monomers by severing filaments and promoting actin disassembly from the pointed (-) end. ATP-actin monomers are re-formed by the action of prolfilin, which promotes exchange of ADP for ATP.

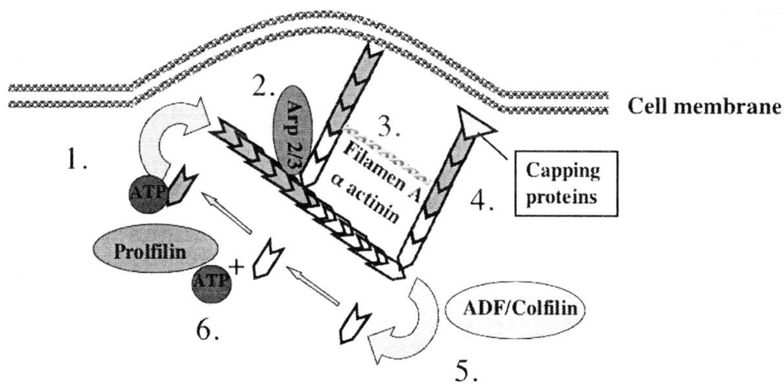

Figure 3: Extension of the leading edge is driven by actin treadmilling. Key steps in this process include: (1) Actin filaments grow by addition of ATP-actin to the barbed (+) end of the filament. (2) Branching of filaments is mediated by activation of the Arp 2/3 complex which binds to the sides or tip of existing filaments and induces the formation of a new daughter filament at a 70^0 angle from the mother filament. (3) These branches are stabilized by cross-linking of the individual filaments by filamen A and α-actinin. (4) Growth of fibers is terminated by the action of capping proteins. (5) Members of the ADF/Cofilin family of proteins regenerate the pool of free actin monomers by severing filaments and promoting actin disassembly from the pointed (-) end. (6) ATP-actin monomers are re-formed by the action of prolfilin, which promotes exchange of ADP for ATP (adapted from (Lodish et al., 2003)).

The rate of actin polymerization, and thus the rate of extension of the leading edge, is regulated by the Rho family of small GTP-binding proteins (Hall, 1998; Raftopoulou and Hall, 2004; Ridley, 2001; Ridley et al., 2003). The activity of Rho family GTPases, in turn is regulated by a number of different proteins including GEFs (Guanine-nucleotide exchange factors), GAPs (GTPase activating proteins) and GDIs (guanine-nucleotide dissociation inhibitors) (Symons and Settleman, 2000). These proteins allow Rho family members to cycle between active, GTP-bound forms, and inactive GDP-bound forms (figure 4A). Rho-family members, Rac1 and Cdc42 promote the assembly of actin at the leading edge (figure 4B). Activated forms of Rac1 and Cdc42 act through a number of effector proteins to increase the rate of actin polymerization and uncapping of actin filaments, while inhibiting actin depolymerization. Cdc42 activates N-WASP (Neural Wiskott-Aldrich syndrome protein). Rac1 activates IRSp53 (insulin receptor substrate p53), which forms a tri-molecular complex with the proteins N-WASP and WAVE (WASP family Verprolin-homologous protein), that activates the Arp 2/3 complex (Miki et al., 2000). Rac1 also activates PIP 5-kinase (phosphitidylinositol-4-phosphate 5-kinase), the enzyme that coverts PIP to PIP2 (phosphitidylinositol-4,5-bisphosphate), leading to barbed-end uncapping and severing of actin filaments (Hartwig et al., 1995). Finally, Cdc42 and Rac1 both activate PAK1 (p21-activated kinase) (Parrini et al., 2002). PAK1 activates LIMK (LIM domain kinase 1), which in turn, phosphorylates cofilin, inhibiting its actin-depolymerizing activity (Edwards et al., 1999; Maekawa et al., 1999).

2. Integrins Regulate Cell Adhesion and Proliferation

Adhesion of cells to the extracellular matrix is mediated by a large family of heterodimeric receptors termed integrins. Integrin receptors consist of one each of 18 α and 8 β subunits, which assemble into at least 24 distinct integrins (Hynes, 2002). Integrins serve as mechanical links between the actin cytoskeleton and sites of extracellular contact, including cell-matrix and cell-cell contacts (reviewed in (Wiesner et al., 2005)). A number of important cellular functions including proliferation, differentiation, apoptosis, and cell migration are regulated by both integrins (Hynes, 2002) and the Rho family of GTPases (Hall, 1998; Ridley, 2001), and key members of the Rho GTPase family including Rho, Rac, and Cdc 42 are mediators of integrin signaling (Miranti and Brugge, 2002). Recent evidence has shown that integrins are essential in targeting Rho family GTPases to the plasma membrane via lipid rafts (del Pozo et al., 2004), as well as their coupling to downstream effectors. Activation of both Rac and Cdc42 requires dissociation of Rho-GDI (see figure 4A, above), and this dissociation is promoted by integrins (Del Pozo et al., 2002).

Binding of integrins to ECM proteins also leads, in a Rac-dependent manner, to the formation of focal complexes containing the membrane-cytoskeletal crosslinking proteins talin, vinculin, paxillin and phosphoproteins such as FAK (focal adhesion kinase) (Hynes, 2002; Zamir and Geiger, 2001). FAK binds to the cytoplasmic tails of β1 integrins, leading to autophosphorylation of FAK on Tyr^{397} and enhanced catalytic activity (Schaller, 2001). FAK binds and phosphorylates the p85 subunit of Pi3- Kinase (Phosphoinositide 3-OH kinase) (Chen and Guan, 1994) and the adapter protein Grb7 (growth factor receptor bound protein 7) (Shen and Guan, 2001). This phosphorylation pathway appears to promote cell migration, while an alternate pathway involving FAK mediated phosphorylation of the adapter protein Grb2 appears to promote cell-proliferation (Shen and Guan, 2001) (figure 5).

Different members of the β1 integrin family have been proposed to play specific roles in promoting migration or proliferation of neuronal stem cells (Jacques et al., 1998; Schmid and Anton, 2003; Tate et al., 2004), with α6β1 integrin acting to promote migration, while αvβ1 and α5β1 integrins were required for cell proliferation (Jacques et al., 1998). α1-, β8, and β1 subunits are expressed in the embryonic and early post -natal RMS, while αv and β6-integrins appear postnatally and persist into adulthood (Murase and Horwitz, 2002). α6β1 integrin is also expressed in the adult RMS, and appears to play a role in directional migration since injection of antibodies to α6β1 integrin disrupted neuronal migration, while injection of the α6β1 ligand, laminin, outside the RMS led to ectopic migration of cells towards the site of infusion (Emsley and Hagg, 2003). α3β1 integrins have also been proposed to play a role in the regulation of neuronal migration through their interaction with reelin and its downstream effector protein Dab1 (Dulabon et al., 2000; Forster et al., 2002; Schmid et al., 2005) (see section below on "Changing Directions").

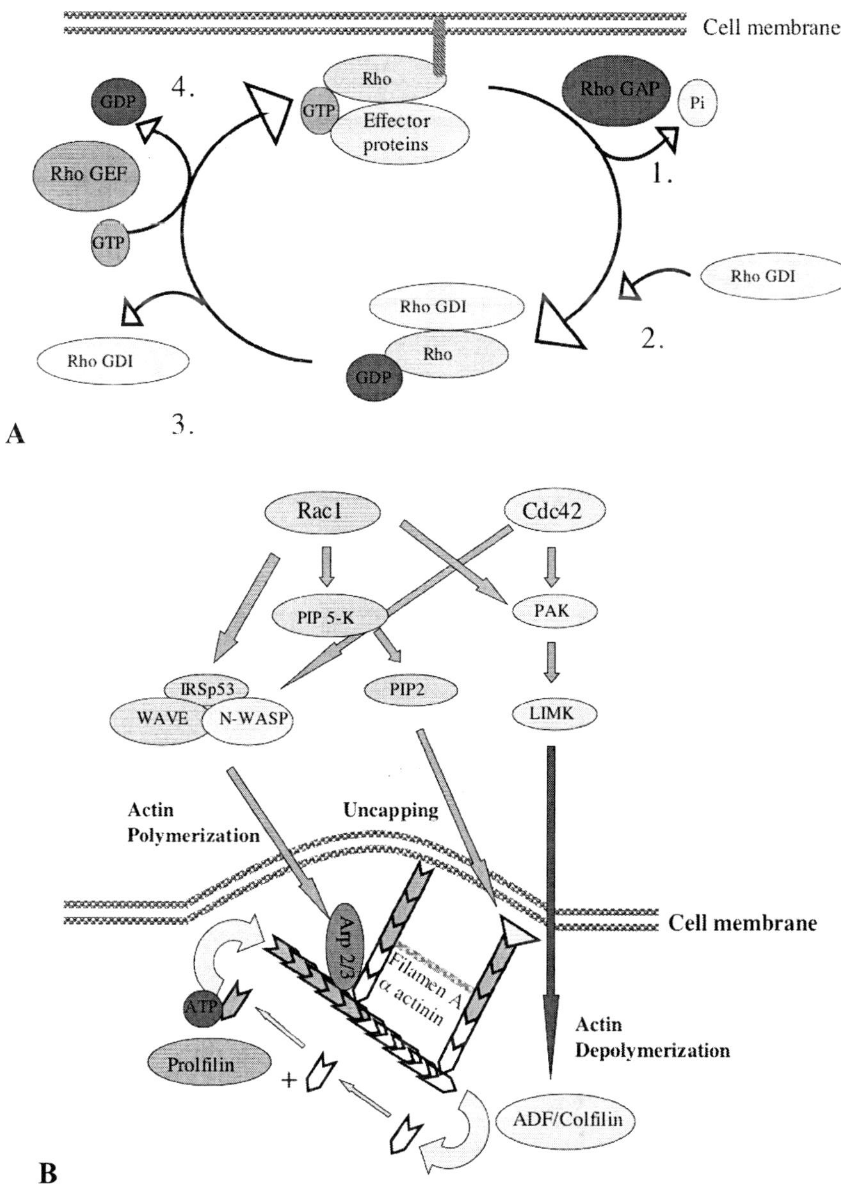

Figure 4: Role of Rho-family GTPases in actin assembly at the leading edge. (A) The activity of Rho-family GTPases is regulated by association with guanine-nucleotides. In their active state Rho family members are bound to GTP and form complexes with specific effector molecules (shown below in B). Binding of active Rho to specific membrane domains is mediated by a C-terminal prenyl group. Rho proteins become inactive following hydrolysis of GTP to GDP. (1) The intrinsic GTPase active of Rho proteins (and thus their rate of inactivation) is increased by proteins termed Rho GAPs (GTPase activating proteins). (2) Rho proteins can be maintained in an inactive, GDP-bound form when they form complexes with Rho GDIs (guanine-nucleotide dissociation inhibitors). (3) Dissociation of Rho GDIs allows Rho proteins to associate with GEFs (Guanine-nucleotide exchange factors), which return them (4) to the active, GTP-bound state by promoting the exchange of GDP for GTP (adapted from (Watanabe et al., 2005)). (B) Rho-family members, Rac1 and Cdc42 promote the assembly of actin at the leading edge. Activated forms of Rac1 and Cdc42 act through a number of effector proteins to increase the rate of actin polymerization and uncapping of actin filaments, while inhibiting actin

depolymerization. Effector proteins include IRSp53 (insulin receptor substrate p53), N-WASP (Neural Wiskott-Aldrich syndrome protein) and WAVE (WASP family Verprolin-homologous protein), which act on the Arp 2/3 complex to promote actin polymerization; PIP 5-kinase (phosphitidylinositol-4-phosphate 5-kinase), the enzyme that coverts PIP to PIP2 (phosphitidylinositol-4,5-bisphosphate), leading to barbed-end uncapping; and PAK1 (p21-activated kinase), which activates LIMK (LIM domain kinase 1), which in turn, phosphorylates cofilin , inhibiting its actin-depolymerizing activity (adapted from (Ridley, 2001)).

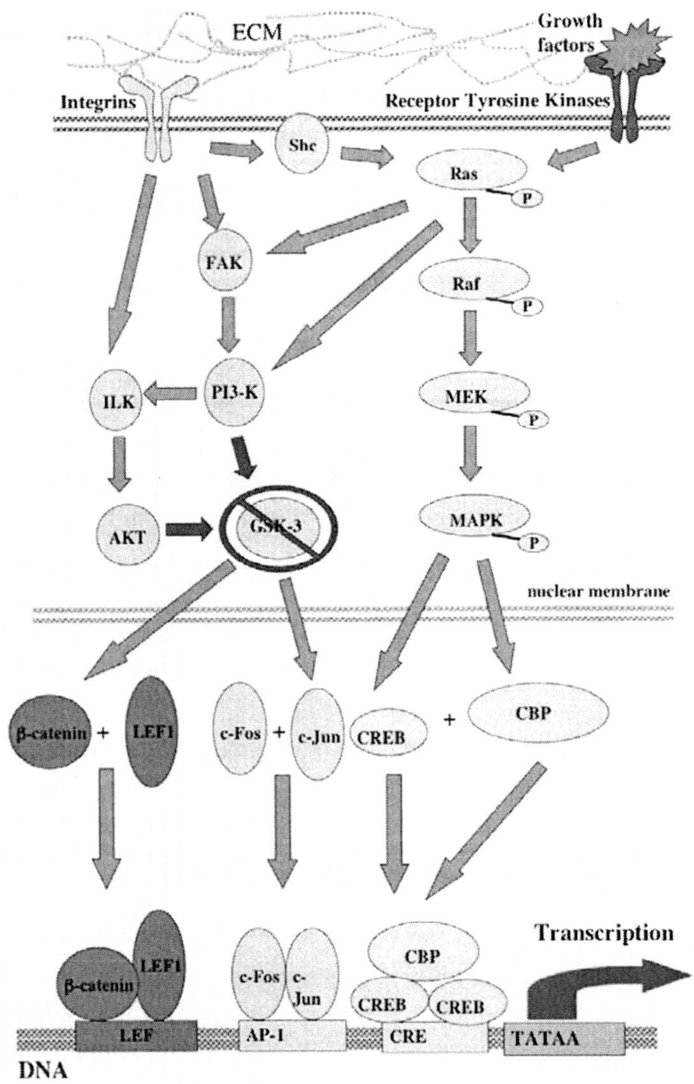

Figure 5: Cross talk between Integrin and Receptor Tyrosine Kinase pathways. Binding of integrins to ECM proteins leads, in a Rac-dependent manner to the formation focal complexes containing the membrane-cytoskeletal crosslinking proteins talin, vinculin, paxillin and phosphoproteins such as FAK (focal adhesion kinase). FAK binds to the cytoplasmic tails of β1 integrins, leading to autophosphorylation on Tyr^{397} and enhanced catalytic activity (Schaller, 2001). FAK binds and phosphorylates the p85 subunit of Pi3- Kinase (Phosphoinositide 3-OH kinase)(Chen and Guan, 1994). This phosphorylation event appears to promote cell migration, while an alternate pathway involving

FAK mediated phosphorylation of the adapter protein Grb2 (growth factor receptor bound protein 2) appears to promote cell-proliferation (Shen and Guan, 2001). Pi3-kinase catalyzes the formation of phosphatidylinositol (3,4,5) triphosphate (PIP3). PIP3 binds directly to ILK and stimulates the phosphorylation of AKT (protein kinase identified in the AKT virus, also known as protein kinase B) and GSK-3 (Glycogen synthase kinase 3) (Delcommenne et al., 1998). Activation of AKT activates a number of pathways inhibiting apoptosis and promoting cellular proliferation (Cantley, 2002; Coffer et al., 1998; Hemmings, 1997; Vanhaesebroeck and Alessi, 2000). Among them, AKT phosphorylates and inactivates GSK-3 (Cross et al., 1995), as well as the pro-apoptotic proteins BAD (bcl2 antagonist of cell death) (Downward, 1998)and caspace 9 (Cardone et al., 1998); cyclin-dependent kinase inhibitor, p27(Liang et al., 2002; Viglietto et al., 2002); and the forkhead transcription factor FKHRL-1 (Brunet et al., 1999) (adapted from (Dedhar, 2000)).

3. Contraction of the Cytoplasm is Regulated by Actinomyosin Contractility and Rho

Actinomyosin contraction generates the force required for contraction of the cytoplasm and forward movement of the cell (Li et al., 2005). Actinomyosin contraction generates a traction force on the substrate at the sites of focal adhesions. This results in a counter-force that pulls the cell forward. Rho acts via ROCKs (Rho-kinases) to regulate actinomyosin contractility (Mitchison and Cramer, 1996; Rafelski and Theriot, 2004; Ridley et al., 2003). ROCKs both directly phosphorylate myosin light chains and phosphorylate and inhibit MLC phosphatase, leading to an increase in contractility (Amano et al., 2000; Kaibuchi et al., 1999). Myosin light chains are also phosphorylated by MLCK (myosin light chain kinase), which is activated by calcium-calmodulin (Stull et al., 1993) and inhibited by PAK (Sanders et al., 1999). MLCK can also be activated by phosphorylation by the MAPKs, ERK1 (extracellular signal regulated kinase) and ERK2, which results in increased MLC phosphorylation and enhanced cell migration (Cheresh et al., 1999; Klemke et al., 1997; Nguyen et al., 1999).

4. Cell Detachment and Recycling of Membrane Receptors are Regulated by RhoA

Cell detachment is mediated in part by internalization of plasma membrane cholesterol and lipid raft associated proteins (del Pozo et al., 2004; Guan, 2004). While Rac1 and Cdc42 regulate the formation of focal adhesions at the front of migrating cells (see above), RhoA is required for the maturation of focal adhesions behind the leading edge and dissociation of focal adhesions at the rear (Kaverina et al., 2002).

Interactions between microtubules and actin also help to drive asymmetric microtubule assembly behavior in cells, such that microtubule growth occurs at the leading edge and microtubule shortening occurs at the cell body and rear (Waterman-Storer and Salmon, 1999). Microtubule growth and shortening, in turn, activate Rac1 and RhoA signaling, respectively, to control actin dynamics (Watanabe et al., 2005). Microtubule stabilization at the leading edge of migrating cells is regulated by Rho and its effector molecule Diaphanous (mDia) (Palazzo et al., 2001). Microtubule targeting of focal adhesions appears to promote the turnover of adhesions in protruding zones to allow further protrusion as well as the release of focal adhesions at retracting cell edges (Small and Kaverina, 2005).

Chain Migration in the RMS: Role of Homotypic Cell-Cell Interactions

The model of cell migration described above is based largely on studies of cultured fibroblasts migrating on a 2- dimensional surface containing one or more protein components of the extra-cellular matrix. Substrates that have been used in these studies include fibronectin, vitronectin, laminin, and collagen. However, it is important to note that cells migrating in a 3- dimensional matrix exhibit different properties form those migrating in 2 dimensions (Li et al., 2005). In particular, it has been suggested that the forces responsible for forward motion of migrating cells in the RMS result, not from cell-matrix, but from homotypic cell-cell interactions (Alvarez-Buylla, 1997; O'Rourke, 1996; Rousselot et al., 1995; Wichterle et al., 1997). These interactions appear to be mediated, at least in part, by the polysialylated form of neural cell adhesion molecule1 (PSA-NCAM) (O'Rourke, 1996). PSA-NCAM is expressed in the subventricular zone and along the length of the RMS (Bonfanti and Theodosis, 1994; Rousselot et al., 1995). Knockout mice, lacking either the entire NCAM1 gene (Cremer et al., 1994) or a specific 180 kilodalton isoform (Tomasiewicz et al., 1993), show abnormalities in neuronal migration, which result in a smaller olfactory bulb and an accumulation of neuronal precursors in the SVZ and proximal RMS. *In vitro* migration of SVZ derived cells can be blocked by enzymatic removal of the polysialic acid moiety from PSA-NCAM (Hu et al., 1996). Migrating cells in the RMS seem to have a similar requirement for PSA-NCAM *in vivo*, since removal of the PSA moieties from NCAM *in vivo* by injection of the enzyme endoneuraminidase N (endo N) into the intracranial cavity results in defects in migration that are similar to those seen in NCAM knockout mice (Ono et al., 1994). This suggests that the polysialic acid moiety is essential for neuronal migration in the RMS. Removal of PSA-NCAM also leads to premature maturation of SVZ-derived cells (Petridis et al., 2004). This effect is dependent on cell-cell contact, which suggests that one function of PSA-NCAM may be to prevent the premature activation of contact-dependent signaling cascades triggered by adhesion molecules.

Migrating cells in the RMS also express neural cell adhesion molecule L1 (L1CAM) (Jenkins et al., 2001). L1CAM is a member of the immunoglobulin superfamily of cell adhesion molecules (CAMs) that mediate cell-to-cell adhesion at the cell surface. L1CAMs mediate both homophilic (L1-L1) and heterophilic binding interactions (reviewed in (Haspel and Grumet, 2003)). L1CAM has been implicated in a large number of important functions in the development of the nervous system including guidance of neurite outgrowth in development, neuronal cell migration, axon bundling, synaptogenesis, myelination, neuronal cell survival, and long-term potentiation. Mutations in the L1CAM gene or deletion of the gene in knockout mice result in an X-linked mental retardation syndrome termed CRASH (corpus callosum agenesis, retardation, aphasia, spastic paraplegia, hydrocephalus)(reviewed in (Kenwrick et al., 2000)). L1CAM potentiates integrin-dependent cell migration on extracellular matrix proteins including fibronectin, vitronectin, and laminin, through a pathway involving c-Src, phosphatidylinositol-3 kinase, and MAP kinase (Thelen et al., 2002). Mutations in L1CAM associated with CRASH lead to loss of both L1-potentiated migration and MAP kinase activation, suggesting that defects in migration may contribute to some of the neurological deficits seen in these patients.

Surprisingly, in view of the fact that both PSA-NCAM and L1CAM both appear to promote cell migration, removal of PSA was found to augment L1-mediated adhesion (Acheson et al., 1991; Fujimoto et al., 2001). Thus, while both PSA-NCAM and L1CAM may contribute to homotypic interactions among migrating cells in the RMS, the nature of those interactions and the relationship of this type of migration to the substrate-dependent models described earlier remain incompletely understood.

Linking Directional Information to the Cytoskeleton

Migrating cells in the SVZ and RMS are guided by mechanisms similar to those used to guide developing axons to their appropriate targets. These mechanisms include chemo-attraction, chemo-repulsion, contact attraction, and contact repulsion (Guan and Rao, 2003; Park et al., 2002; Tessier-Lavigne and Goodman, 1996). In addition, activity-dependent mechanisms may play a role in determining the phenotype of newly generated prospective neurons, even while they are in the process of migrating and dividing along the course of the RMS (Bolteus and Bordey, 2004; Hack et al., 2005; Liu et al., 2005).

Factors which may play a role in directional migration in the RMS include netrins (Murase and Horwitz, 2002), semaphorins (Bagri and Tessier-Lavigne, 2002; He et al., 2002; Tamagnone and Comoglio, 2004), slits (Mason et al., 2001; Nguyen-Ba-Charvet et al., 2004; Wu et al., 1999), ephrins (Conover et al., 2000), and neuregulins (Anton et al., 2004). Netrins, slits and some semaphorins are secreted molecules that associate with cells or the extracellular matrix, while neuregulins, ephrins and other semaphorins are expressed on the cell surface (Falls, 2003; Yu and Bargmann, 2001). Each of these molecules interacts in various ways with members of the Rho GTPase family described in the first section. These interactions are thought to mediate the effects of the various chemoattractant and chemorepulsive factors of the actin cytoskeleton, and this in turn determines the effects of these factors on the cell's migration. Some of these interactions are described in more detail below.

Netrins

Netrins are large (approximately 70-80 kDa), soluble proteins that are closely related to the extracellular matrix molecule laminin (Tessier-Lavigne and Goodman, 1996; Yu and Bargmann, 2001). Netrins were originally characterized as chemo-attractants for spinal commissural axons during embryonic development (Serafini et al., 1996). However, subsequent work showed that the same netrin can be attractive to some axons and repulsive to others depending on which netrin receptors they express. Vertebrate receptors for netrins include Deleted in Colorectal Cancer (DCC) (Keino-Masu et al., 1996), neogenin (Yu and Bargmann, 2001), rostral cerebellar malformation (rcm) (Ackerman et al., 1997) and UNC-5H-1, -2, -3, and -4 (Engelkamp, 2002; Leonardo et al., 1997; Przyborski et al., 1998). Migrating cells in the adult RMS express both DCC and neogenin, and olfactory mitral cells (which are the synaptic targets of the newly generated granule and periglomerular cells in the bulb (Shepard and Greer, 1998)) express netrin-1 (Murase and Horwitz, 2002). *In vitro* experiments have shown that netrin-1 repels SVZa derived neuroblasts (Liu and Rao, 2003). However, interactions between netrins and DCC may have additional roles in regulating migration in the RMS, since addition of antibodies to DCC to slice cultures led migrating cells to move in a non-directed manner including back towards the SVZa or at right angles to the normal migratory pathway (Murase and Horwitz, 2002).

How is directional information from netrin-DCC interactions conveyed to the cytoskeleton? Netrin binding to DCC activates FAK and the tyrosine kinases Fyn and Src, stimulating tyrosine phosphorylation of these proteins as well as DCC itself (Li et al., 2004; Liu et al., 2004; Ren et al., 2004). Netrins also activate Rac and Cdc42 (Li et al., 2002; Shekarabi and Kennedy, 2002). As described above, the actions of these two proteins stimulate actin polymerization at the leading edge of migrating cells. While the majority of the work on signaling cascades downstream of netrin signaling has been done in the context of axonal guidance, it is likely that similar mechanisms underlie the requirement for DCC in directional movement in the RMS.

Semaphorins

Semaphorins are a family of highly conserved molecular signaling molecules that, like the netrins, were originally identified through their roles in axon guidance. Subsequent work showed that semaphorins are also involved in a number of other cellular functions including cell migration (reviewed in (Tamagnone and Comoglio, 2000; Tamagnone and Comoglio, 2004)). In contrast to netrins, semaphorins act primarily as chemo-repellants (Kolodkin et al., 1993; Pasterkamp and Kolodkin, 2003). Semaphorins include both membrane-associated and secreted molecules. All semaphorin family members have a secretion signal, a 500-amino acid sema domain, and 16 conserved cysteine residues. Semaphorin receptors include two families of proteins: the plexins and the neuropilins (Fujisawa and Kitsukawa, 1998). Membrane-bound semaphorins bind directly to plexins, while secreted semaphorins require neuropilins as co-receptors (Tamagnone and Comoglio, 2004). Semaphorin receptors can also form complexes with other transmembrane molecules including L1CAM (discussed above).

While there is currently no direct evidence for semaphorin-mediated signaling in directional guidance in the adult RMS, semaphorins are known to play key roles in the tangential migration of GABAergic interneurons during embryonic development (Marin et al., 2001; Tamamaki et al., 2003). Migrating interneurons expressing neuropilins are directed to the cortex, while those lacking them go to the striatum. This effect is mediated by a repulsive signal from the striatum, composed at least in part by Sema3A and Sema3F (Marin et al., 2001; Tamamaki et al., 2003). Sema3A is also expressed by mitral and tufted cells of the adult olfactory bulb (Giger et al., 1998) and numerous other members of the semaphorin family continue to be expressed in the adult brain (reviewed in (de Wit and Verhaagen, 2003)).

Many of signaling pathways implicated in cell migration can be regulated by semaphorins (Guan and Rao, 2003; Negishi et al., 2005; Tamagnone and Comoglio, 2004). Sema3A and Sema4D have been reported to inhibit both integrin-mediated cell adhesion and migration of endothelial cells (Barberis et al., 2004; Serini et al., 2003). Plexin activation in adherent cells leads to retraction of cellular processes and cell rounding in a process analogous to semaphorin-mediated "growth cone collapse" (Takahashi et al., 1999). This cellular collapse involves the disassembly of integrin-based focal adhesions as well as actin depolymerization.

Sema3A-induced growth cone collapse requires the activity of Rac1 (Jin and Strittmatter, 1997; Kuhn et al., 1999). The requirement for Rac1 activity in Sema3A-induced repulsive activity was a surprise since, as described above, Rac1 activity is usually associated with actin polymerization and extension of the leading edge. Rac1 is also required for Sema3A-induced endocytosis of growth cone plasma membranes and re-organization of F-actin during growth cone collapse, which may explain the requirement for Rac1 activity in Sema3A-mediated repulsion (Jurney et al., 2002).

Semaphorin signaling also involves activation of Rho. Plexin-B1 and Plexin-B2 activate Rho through two specific Rho-GEFs, PDZ-Rho-GEF and leukemia-associated Rho-GEF (LARG) (Perrot et al., 2002; Swiercz et al., 2002). Activation of these GEFs causes Rho to exchange GDP for GTP and leads to activation of Rho and Rho kinase. As described above, activation of Rho leads to cell detachment and recycling of membrane receptors, two processes that may be important in the repulsive effects of semaphorins on migrating cells.

Slit/Robo Proteins

Slits are secreted proteins, originally identified in Drosophila in a screen for genes involved in pattern formation, and subsequently shown to act as chemo-repellents for developing axons (Wong et al., 2002). Slits repel neuronal precursors migrating from the SVZa to the olfactory bulb (Wu et al., 1999), as well as precursors of GABAergic interneurons migrating from the lateral ganglionic eminence (LGE) to the neocortex (Zhu et al., 1999). Three slit genes have been cloned in mammals, Slit1, Slit 2, and Slit 3. Slit 1 is expressed primarily in brain, while Slit 2 is expressed primarily in the spinal cord and Slit3 is expressed in the thyroid gland (Itoh et al., 1998). Roundabouts (Robos) are cell surface receptors that are responsible for the repulsive effect Slit proteins in axon guidance (Brose et al., 1999; Long et al., 2004). Activation of Robo receptors by slit proteins inhibits N-cadherin mediated cell adhesion (Rhee et al., 2002) and can also silence netrin-mediated attractive signals (Stein and Tessier-Lavigne, 2001).

Slit1 and slit2 are expressed in the septum (Hu, 1999; Wu et al., 1999), while Robo2 and Robo3 are expressed by migrating cells in the RMS (Marillat et al., 2002). Both the septum and the choroid plexus exert endogenous chemo-repulsive activities on migrating SVZa derived neuroblasts (Hu, 1999; Wu et al., 1999). This activity consists of a combination of slit1 and slit2 (Nguyen-Ba-Charvet et al., 2004). A novel family of Rho GAPs, slit-robo

GAPS (srGAPs), has been described that interacts selectively with the intracellular domain of Robo proteins (Wong et al., 2001). srGAP1 binds to and inhibits both Cdc42 and RhoA, but not Rac1. The effect of slits on migrating SVZa derived cells is lost in cells expressing a constitutively active form of Cdc42, indicating that down regulation of Cdc42 is crucial to the repulsive effects of slits on migrating neuroblasts (Wong et al., 2001). Surprisingly, slit1 is also expressed by the migrating neuroblasts themselves, as well transit amplifying cells, suggesting that the regulation of migration by slit-robo interactions is more complicated than simple repulsion from non-target tissues (Nguyen-Ba-Charvet et al., 2004).

Ephrins

Ephrins are ligands for the eph family of tyrosine kinases. Ephrins and ephs are involved in a number of important aspects of neural development including axon guidance, neural crest cell migration, and establishment of segmental boundaries in the developing nervous system. (Flanagan and Vanderhaeghen, 1998; O'Leary and Wilkinson, 1999; Palmer and Klein, 2003; Wilkinson, 2001). Ephrins and their receptors are divided into two classes based on their mode of attachment to the cell membrane and binding specificities. Ephrin-A ligands are anchored to the cell membrane by glycosylphosphatidylinositol (GPI) linkage, and bind to EphA receptors. Ephrin-B ligands are transmembrane proteins that bind EphB receptors. The EphA4 receptor binds to both classes of ephrin ligands. A striking feature of ephrin/eph signaling is the bi-directional nature of the signal. Ephrin binding induces dimerization of the eph receptor and tyrosine phosphorylation of the receptor itself as well as downstream substrates. In the case of ephrin-B/ephB binding, it also increases tyrosine phosphorylation of the intracellular domain of ephrin B. This phosphorylation is mediated by the Src family of tyrosine kinases (Palmer et al., 2002). Ephrin-A/ephA signaling also appears to be bi-directional and leads to an increase in integrin-mediated cell adhesion. This process requires the Fyn tyrosine kinase and a 120-kDa raft membrane protein that is tyrosine-phosphorylated specifically after ephrin-A activation (Davy et al., 1999; Davy and Robbins, 2000; Huai and Drescher, 2001).

Ephrin-B2 and ephrin-B3 are expressed by glial tube astrocytes in the SVZ and RMS, while EphA4, EphB2, and EphB3 are expressed by as yet unidentified cells in the RMS (Conover et al., 2000). Infusion of the extracellular portion of either EphB2 or ephrin-B2 into the lateral ventricles disrupts migration of neuronal precursors in the RMS and leads to an increase in the proliferation of SVZ astrocytes (Conover et al., 2000). As noted in the introduction, SVZ astrocytes serve both as the precursors of, and a physical channel for, the migrating cells (Alvarez-Buylla and Garcia-Verdugo, 2002; Doetsch et al., 1999). The fact that infusion of extracellular portions of either EphB2 or ephrin-B2 leads to a similar result suggests that bi-directional signaling between SVZ astrocytes and migrating neuroblasts may be important in regulating the balance between proliferation and migration of newly generated cells in the adult SVZ. The bi-directional nature of this signal may relate to the requirement for glial cells in maintaining an appropriate "niche" for adult neurogenesis (Lim and Alvarez-Buylla, 1999; Song et al., 2002).

Like the other proteins discussed above, eph/ephrin signaling is also coupled to cytoskeletal changes by the Rho family of GTPases. Eph/ephrin signaling activates a variety of exchange factors that regulate Rho family GTPases in neurons (reviewed in (Murai and Pasquale, 2005; Noren and Pasquale, 2004)). These exchange factors include ephexin1 (Sahin et al., 2005; Shamah et al., 2001), vav2 (Cowan et al., 2005), intersectin-1 (Irie and Yamaguchi, 2002), kalirin-7 (Penzes et al., 2003), and tiam1 (Tanaka et al., 2004). Ephexin1 is coupled to EphA receptors, while intersectin-1 and kalirin-7 are coupled to EphB, and tiam1 and vav2 can be coupled to either EphA or EphB (Murai and Pasquale, 2005).

The activation of EphA receptors induces the activation of RhoA and the inhibition of Rac1 (Wahl et al., 2000). These effects are mediated by ephexin1, a GEF for Rho that interacts directly with the kinase domain of EphA receptors. Stimulation of EphA receptors by Ephrin-A leads to tyrosine phosphorylation of ephexin1 and increases the GEF the activity of ephexin1 leading to RhoA activation (Sahin et al., 2005). Other results suggest that Rac1 is required for EphA mediated effects on endocytosis of the growth cone plasma membrane and reorganization of F-actin (Jurney et al., 2002). Thus, both RhoA and Rac1 may be required for ephrin-mediated effects on cell migration.

Vav exchange factors are also known to activate Rho GTPases (Schmidt and Hall, 2002), and tiam1 has been shown to activate Rac1 downstream of ephrinB1 (Tanaka et al., 2004). As noted above, Rac1 is thought to play an important role in regulating extension of the leading edge in migrating cells, while Rho plays important roles in regulating both cell contraction and cell detachment. However, roles for specific GEFs have not yet been identified in relation to the effects of EphrinB2 and EphB2 on cell migration and proliferation in the adult SVZ and RMS.

Neuregulins

The neuregulins (NRGs) are cell-cell signaling proteins that are ligands for receptor tyrosine kinases of the ErbB family (reviewed in (Buonanno and Fischbach, 2001; Carpenter, 2003; Falls, 2003; Riese and Stern, 1998; Yarden and Sliwkowski, 2001)). ErbB4 is expressed in migrating neuroblasts in the SVZ and RMS of adult mice (Anton et al., 2004), as well as other populations of neuronal precursors that undergo tangential migration (Flames et al., 2004; Yau et al., 2003). ErbB4 can be activated by a large number of ligands including neuregulins (NRG1- NRG4), betacellulin, HB-EGF, and epiregulin. Following ligand binding, the ErbB4-ectodomain is cleaved by the metalloprotease TACE (tumor necrosis factor-alpha converting enzyme) (Rio et al., 2000; Zhou and Carpenter, 2000). A subsequent cleavage by gamma-secretase releases the ErbB4 intracellular domain from the membrane and facilitates its translocation to the nucleus, where it may influence gene transcription (Carpenter, 2003; Ni et al., 2001).

NRG1 is expressed in the SVZ and RMS, and *in vitro* experiments showed that NRG1 acts as a preferred migration substrate for neuroblasts derived from normal mice, but not those derived from mice with inducible mutations in the ErbB4 gene (Anton et al., 2004). Genetic linkage studies have implicated the NRG1 gene as a candidate gene for risk for schizophrenia (Arnold et al., 2005; Stefansson et al., 2003; Stefansson et al., 2002; Zhao et al., 2004), and mutant mice heterozygous for either NRG1 or its receptor, ErbB4, show a behavioral phenotype that overlaps with mouse models for schizophrenia (Stefansson et al., 2002). ErbB4 mutant mice also showed deficits in chain migration characterized by a shift from chain migration to migration as single cells, and a slower rate of migration and loss of directionality in slice culture. NRG1 also as acts as an attractant for GABAergic cortical interneurons undergoing tangential migration (Flames et al., 2004). These observations suggest that abnormal neuronal migration that results from defects in NRG1/ErbB4 signaling may contribute to the risk of developing schizophrenia.

NRG1 stimulation has been shown to induce the formation of a multiprotein complex that contains the tyrosine kinase Pyk2 as well as p190 RhoGAP (p190), RasGAP, c-Abl and the focal adhesion molecules p130cas and paxillin (Zrihan-Licht et al., 2004). This is consistent with previous work showing that NRG1 can regulate cytoskeletal reorganization and cell migration through PAK1 and Pi3-K (Adam et al., 1998). As discussed above, Pi3-K is downstream of FAK and paxillin in the integrin signaling pathway, and PAK1 can activate LIMK, which in turn, phosphorylates cofilin, inhibiting its actin-depolymerizing activity and thus promoting cell migration (Edwards et al., 1999; Maekawa et al., 1999).

Changing Directions: The Transition from Tangential to Radial Migration

The activities of the various signaling molecules discussed above including netrins, semaphorins, slits, ephrins, and neuregulins, all relate to the tangential migration of cells from the SVZ to the olfactory bulb along the RMS. However, before integrating into the synaptic architecture of the bulb, the migrating cells must leave the RMS and migrate to the appropriate cell layers in the olfactory bulb where they then differentiate into two types of interneurons: granule cells and periglomerular cells (Lois and Alvarez-Buylla, 1994; Luskin, 1993). Two molecules in the extracellular matrix, reelin (Hack et al., 2002) and Tenascin-R (Saghatelyan et al., 2004), as well as the secreted factor prokineticin-2 (Ng et al., 2005) have been proposed to key roles in the transition from radial to tangential migration. Below, I discuss what is known about each of these molecules as it relates to the transition from tangential to radial migration in the adult RMS.

Reelin is a large extra-cellular matrix protein that is important in the formation of normal layered structures in the brain during embryonic development (Lambert de Rouvroit and Goffinet, 1998; Rugarli and Ballabio, 1995). Mice lacking both copies of the reelin gene develop "inside-out" brains, with a reversal of the normal layers of the cerebral cortex (Caviness and Sidman, 1972; Rakic and Caviness, 1995; Rice and Curran, 2001). Reelin mRNA (Alcantara et al., 1998) and protein (Pappas et al., 2003) are expressed in the adult olfactory bulb, and reeler mice exhibit abnormalities in the architecture of the olfactory bulb similar to those seen in the cerebral cortex (Caviness and Sidman, 1972). The reelin protein has been proposed to serve as a detachment signal during the migration of neuronal precursors from the subventricular zone to the olfactory bulb in the adult brain (Hack et al., 2002). Immuno-labeling for the reelin protein in heterozygous mice shows an accumulation of reelin-positive cells in the rostral migratory stream that are apparently unable to shift from tangential to radial migration (Pappas et al., 2003). Reeler heterozygous mice also exhibit deficits in olfactory learning (Larson et al., 2003), suggesting that this abnormal migration has functional consequences in the adult brain.

Reelin receptors include the very low density lipoprotein receptor (VLDLR) and the apoE receptor 2 (ApoER2) (Hiesberger et al., 1999). Binding of reelin to these receptors induces tyrosine phosphorylation of the cytosolic adaptor protein, Disabled-1 (Dab1). "Scrambler" (Ware et al., 1997) or "yotari" (Sheldon et al., 1997) mice have mutations in the Dab1 gene and have a phenotype identical to that of the reeler mouse, demonstrating that the tyrosine phosphorylation of Dab1 mediates the biological effects of reelin signaling. *In vitro* experiments in a Matrigel culture system (Hack et al., 2002) showed that reelin acted as a detachment signal for chain-migrating interneurons precursors in the olfactory bulb, inducing their dispersal into individual cells.

As discussed above, reelin has been shown to associate with α3β1 integrins. Experiments using α3β1 integrin-deficient mouse embryos showed that effects of reelin on cortical neuronal migration were dependent on interactions between reelin and α3β1 integrins (Dulabon et al., 2000). A similar requirement for reelin-integrin interactions has been shown in the formation of the radial glial scaffold in the hippocampus (Forster et al., 2002). These authors also showed that recombinant Reelin induced branching of glial processes in vitro, suggesting that reelin may play a role in regulating interactions between migrating neurons and radial glia. Subsequent work has shown that α3β1 integrins binds to the N-terminal region of reelin, a site distinct from the region of reelin shown to associate with VLDLR and ApoER2, while Dab1binds to the cytoplasmic region of β1 integrins in a reelin-dependent manner (Schmid et al., 2005). These observations suggest that α3β1 integrins mediate the effects of reelin in the transition from tangential to radial migration independently of other reelin receptors such as VLDLR and ApoER2.

Tenascin-R (TNR) is a member of the tenascin gene family (reviewed in (Jones and Jones, 2000; Pesheva and Probstmeier, 2000)). TNR isoforms are generated by pre-mRNA alternative splicing. In the adult mouse TNR immunostaining is restricted to the granule cells and internal plexiform layers of the olfactory bulb (Pesheva and Probstmeier, 2000). Expression of TNR is activity dependent and is markedly reduced by odor deprivation (Saghatelyan et al., 2004). Grafting of TNR-transfected cells into non-neurogenic regions rerouted migrating neuroblasts toward these regions, while in TNR deficient mice neuroblasts were retained in the anterior segment of the rostral migratory stream (Saghatelyan et al., 2004). These observations suggest that tenascin-R may mediate both the detachment of migrating neuroblasts from chains and the initiation of radial migration.

Experiments in other contexts (reviewed in (Pesheva and Probstmeier, 2000)) show that TNR can either enhance or inhibit cell-matrix and neuronal-glial interactions. These effects are both substrate-dependent and isoform-specific. The specific molecules linking TNR to cytoskeletal changes in migrating cells have not yet been determined, but they may involve effects on β1 integrins and downstream phosphorylation of FAK since in other contexts TNR can inhibit both β1 integrin-dependent cell adhesion and tyrosine phosphorylation of FAK (Pesheva and Probstmeier, 2000).

Prokineticin-2 (PK2) is a cysteine rich secreted protein that, along with the related protein prokineticin-1, has been implicated a wide variety of biological functions including contraction of gastrointestinal smooth muscle (Li et al., 2001), regulation of circadian rhythms (Cheng et al., 2002), and regulation of proliferation, survival and migration of endothelial cells (LeCouter et al., 2003). Prokineticins are thought to exert their biological effects by activation of two highly related G-protein coupled receptors (Soga et al., 2002). Activation of prokineticin receptors leads to mobilization of calcium, stimulation of phosphoinositide turnover, and activation of p44/p42 MAPK signaling pathways (Lin et al., 2002).

PK2 is expressed in both periglomerular and granule cells, the two types of adult generated neurons in the olfactory bulb, while its receptors, PKR1 and PKR2, are expressed in the underlying RMS (Ng et al., 2005). PK2 knockout mice have smaller olfactory bulbs, and neuroblasts in these animals fail to make the transition from tangential to radial migration. *In vitro* PKs act as chemoattractants for SVZa derived neuroblasts, and also lead to the dispersal of migrating chains of neuroblasts (Ng et al., 2005). The mechanisms linking PK2 signaling to changes in cytoskeletal dynamics are currently unknown, as are the relationships among PK2, reelin and TNR, all of which have been implicated in the transition from tangential to radial migration in the adult RMS.

Balancing Proliferation and Migration

Unlike the neuronal precursors that give rise to cortical neurons during embryonic development, which divide in the ventricular zone and then migrate to appropriate positions in the overlying cortex, neuronal precursors in the RMS continue to divide as they migrate (Menezes et al., 1995; Rousselot et al., 1995). Maintaining an appropriate balance between proliferation and migration necessitates a series of signaling pathways which co-ordinate these activities. Experiments in a variety of systems have shown that the processes of migration and cell-cycle exit can be coupled (Choi et al., 2005; Conover et al., 2000; Liu et al., 2005). As discussed above, infusion of the extracellular portion of either EphB2 or ephrin-B2 into the lateral ventricles both disrupts migration of neuronal precursors in the RMS and leads to an increase in the proliferation of SVZ astrocytes (Conover et al., 2000). This suggests that bi-directional signaling between SVZ astrocytes and migrating neuroblasts is important in maintaining the balance between migration and proliferation.

Recent work has shown that both the speed of migration of SVZa derived neuroblasts (Bolteus and Bordey, 2004), and the proliferation of SVZ astrocytes (Liu et al., 2005) is regulated by non-synaptic GABA released by migrating neuroblasts. SVZ progenitor cells express the GABA transporter (GAT) (Wang et al., 2003), and uptake of GABA by this transporter is thought to increase the speed of migration by relieving the inhibitory effects of extra-synaptic GABA on the migrating neuroblasts (Bolteus and Bordey, 2004). In addition, depolarization of SVZ precursors leads to a decrease in the rate of proliferation (Liu et al., 2005). These results underscore the importance of bi-directional signaling mechanisms in regulating both proliferation and migration in the SVZ and RMS.

Recent data from our laboratory (Cleary et al., unpublished data) indicate that SVZ astrocytes express two proteins, ezrin and beta-catenin, that may be important in this process. Ezrin is a member of the highly conserved ERM (Ezrin-Radixin-Moesin), band 4.1 family of membrane-cytoskeletal linking proteins (Gould et al., 1989; Takeuchi et al., 1994). ERM proteins are involved in a wide variety of cellular functions including cell motility, signal transduction, cell-cell and cell-matrix recognition (Bretscher et al., 1997) (Bretscher et al., 2002; Louvet-Vallee, 2000). Ezrin is expressed in radial glia (Johnson et al., 2002), and co-localizes with L1CAM in migrating cortical neurons (Mintz et al., 2003). Ezrin binds the intracellular domain of L1CAM (Dickson et al., 2002). This interaction is required for L1CAM-mediated homotypic interactions (Cheng et al., 2005a). As discussed L1CAM-mediated homotypic interactions may play a role in regulating migration in the RMS. Ezrin also binds to Rho-GDI (Hirao et al., 1996; Maeda et al., 1999; Takahashi et al., 1997). Binding of Ezrin to Rho-GDI allows Rho to dissociate from Rho-GDI and become active. In a positive feedback loop, ezrin activity is regulated by phosphorylation by Rho-kinase (reviewed in (Bretscher et al., 2000; Bretscher et al., 2002; Gautreau et al., 2002)).

β-Catenin, in association with the transcription factor LEF1 (Porfiri et al., 1997), is a key downstream mediator of the wnt signaling pathway (Logan and Nusse, 2004). Wnt signaling inhibits neural differentiation of embryonic stem cells by up-regulating the expression bone morphogenetic proteins (BMPs) (Haegele et al., 2003). Endogenous β-catenin is highly expressed in embryonic neural precursors, and transgenic mice that express high levels of a stabilized β-catenin transgene develop enlarged brains with expanded precursor populations (Chenn and Walsh, 2002). Adult mice expressing the same β-catenin transgene develop enlarged forebrains with expanded subventricular zones (Chenn and Walsh, 2003). These results suggest that nuclear β-catenin can promote cell proliferation and inhibit differentiation of stem cells in the adult SVZ. In other cell types ezrin forms a complex with E-cadherin and β-catenin that appears to regulate both cell-cell and cell-matrix recognition (Hiscox and Jiang, 1999). Disruption of this complex by tyrosine phosphorylation of ezrin leads to loss of cell-cell adhesion and increased invasiveness. Decreased expression of E-cadherin in breast cancer cells is also associated with an increase in nuclear β-catenin, a loss of cell-cell adhesion, and an increased rate of cell division (Yang et al., 2001). These observations suggest that sequestration of β-catenin in membrane-associated complexes can serve to regulate the rate of cell division.

Cadherins are calcium-dependent homophilic adhesion molecules that, together with their intracellular binding partners the catenins, are thought to mediate cell-cell adhesion and migration (reviewed in (Nelson and Nusse, 2004; Redies, 2000)). Unlike most cadherins, which show highly restricted patterns of expression during development, N-cadherin is ubiquitously expressed in the entire proliferative neuroepithelium. N-cadherin and β-catenin are co-localized in axons and dendrites prior to synapse formation (Benson and Tanaka, 1998). A complex containing N-cadherin and β-catenin has also been suggested to play a role in down-regulating cell proliferation and inducing terminal differentiation in myoblasts (Gavard et al., 2004). The fact that over-expressing β-catenin leads to increased proliferation in the adult SVZ (Chenn and Walsh, 2003) suggests that the ratio of nuclear to membrane-associated β-catenin may be important in balancing ongoing proliferation and migration in the adult SVZ and RMS as well.

Migrating cells in the RMS also express p120 catenin (Chauvet et al., 2003). Like β-catenin, p120 catenin can bind to the intracellular portion of both E-cadherin and N-cadherin, as well as to β-catenin (Chauvet et al., 2003; Daniel and Reynolds, 1995; Reynolds et al.,

1994; Shibamoto et al., 1995; Staddon et al., 1995). p120 catenin has also been shown to inhibit RhoA (Anastasiadis et al., 2000). Moreover, inhibition of RhoA and binding to cadherins was mutually exclusive, which would allow RhoA to become activated specifically at nascent cell-cell contacts. As discussed above, RhoA is involved in multiple aspects of cell migration including regulation of actinomyosin contractility as well as cell detachment and recycling of membrane receptors. Disruption of stress fibers and focal adhesions by p120 catenin can be blocked by dominant negative Cdc42 and Rac1 or by constitutively active Rho-kinase, but is enhanced by dominant negative RhoA (Grosheva et al., 2001; Noren et al., 2000). These effects are mediated by binding and activation of the RhoGEF, vav2 (Noren et al., 2000) (discussed above, in relation to eph/ephrin signaling). Co-transfection with an excess of E-cadherin leads to sequestration of exogenous p120ctn to cell-cell junctions or to small cadherin-containing vesicles, and abolishes p120ctn effects on cell morphology (Grosheva et al., 2001). Thus, interactions among cadherins and catenins, and perhaps other cytoskeletal crosslinking proteins that effect the subcellular localization of these proteins, appear to be mechanisms by which both proliferation and migration in the SVZ and RMS may be regulated.

Conclusion

Neuronal precursors of adult generated olfactory bulb interneurons migrate long distances first tangentially through the rostral migratory stream, and then radially into the outer layers of the olfactory bulb where they differentiate into two types of inhibitory interneurons: granule cells, and periglomerular cells. This migration requires an ongoing rearrangement of the cytoskeleton as well as integration of chemo-attractive and repulsive cues from the environment to result in directed movement of the cell along the appropriate pathways. Cell migration can be thought of as a cyclical process. Essential steps in this process include: (1) extension of the leading edge; (2) adhesion to matrix contacts; (3) contraction of the cytoplasm; (4) release from contact sites and recycling of membrane receptors from the rear to the front of the cell. Each of these phases, in turn, is dependent upon multiple cyclical biochemical processes, many of which are regulated by members of the Rho family of GTPases. Directional cues present in the environment of the migrating cells exert many of their effects via these same GTPases. Netrins, semaphorins, slits, ephrins, and neuregulins are among the molecules proposed to regulate the directional movement of SVZa derived neuroblasts along the RMS. Each of these families of signaling molecules has been shown to interact with one or more of the Rho family GTPases to bring about their effects on cell migration.

Two molecules in the extracellular matrix, reelin and TNR, and the secreted protein, PK2, have been shown to key roles in the transition from radial to tangential migration. Reelin is thought to exert its effects through interactions with $\alpha3\beta1$ integrins. The mechanism by which TNR exerts its effects is unknown, but may involve interactions with $\beta1$ integrins as well. PK2 signals through two G-coupled receptors that activate the MAPK pathway. However, the mechanisms linking PK2 to cytoskeletal changes have not yet been described.

A central theme of the work described here is that signaling molecules that regulate both cell proliferation and migration do so, in large part, via their effects on the cytoskeleton. Mechanisms involved in regulating the balance between proliferation and migration in include bi-directional signaling between SVZ astrocytes and migrating neuroblasts mediated by ephs and ephrins, and GABAergic signaling from neuroblasts to astrocytes. Membrane-cytoskeletal crosslinking proteins including members of the ERM, catenin and cadherin families, may also play important roles in balancing proliferation and migration of newly generated cells in the adult SVZ and RMS. Understanding the links between specific signaling events and the structural changes they induce in stem cells and their progeny represents one of the major challenges for the next decade. This understanding will help us as we attempt to harness the restorative capacity inherent in stem cells for therapeutic purposes in the CNS and elsewhere in the body.

REFERENCES

Acheson, A., Sunshine, J. L., and Rutishauser, U. (1991). NCAM polysialic acid can regulate both cell-cell and cell-substrate interactions. *J Cell Biol 114*, 143-153.

Ackerman, S. L., Kozak, L. P., Przyborski, S. A., Rund, L. A., Boyer, B. B., and Knowles, B. B. (1997). The mouse rostral cerebellar malformation gene encodes an UNC-5-like protein. *Nature 386*, 838-842.

Adam, L., Vadlamudi, R., Kondapaka, S. B., Chernoff, J., Mendelsohn, J., and Kumar, R. (1998). Heregulin regulates cytoskeletal reorganization and cell migration through the p21-activated kinase-1 via phosphatidylinositol-3 kinase. *J Biol Chem 273*, 28238-28246.

Alcantara, S., Ruiz, M., D'Arcangelo, G., Ezan, F., de Lecea, L., Curran, T., Sotelo, C., and Soriano, E. (1998). Regional and cellular patterns of reelin mRNA expression in the forebrain of the developing and adult mouse. *J Neurosci 18*, 7779-7799.

Alvarez-Buylla, A. (1997). Mechanism of migration of olfactory bulb interneurons. *Semin Cell Dev Biol 8*, 207-213.

Alvarez-Buylla, A., and Garcia-Verdugo, J. M. (2002). Neurogenesis in adult subventricular zone. *J Neurosci 22*, 629-634.

Amano, M., Fukata, Y., and Kaibuchi, K. (2000). Regulation and functions of Rho-associated kinase. *Exp Cell Res 261*, 44-51.

Anastasiadis, P. Z., Moon, S. Y., Thoreson, M. A., Mariner, D. J., Crawford, H. C., Zheng, Y., and Reynolds, A. B. (2000). Inhibition of RhoA by p120 catenin. *Nat Cell Biol 2*, 637-644.

Anton, E. S., Ghashghaei, H. T., Weber, J. L., McCann, C., Fischer, T. M., Cheung, I. D., Gassmann, M., Messing, A., Klein, R., Schwab, M. H., et al. (2004). Receptor tyrosine kinase ErbB4 modulates neuroblast migration and placement in the adult forebrain. *Nat Neurosci 7*, 1319-1328.

Arnold, S. E., Talbot, K., and Hahn, C. G. (2005). Neurodevelopment, neuroplasticity, and new genes for schizophrenia. *Prog Brain Res 147*, 319-345.

Bagri, A., and Tessier-Lavigne, M. (2002). Neuropilins as Semaphorin receptors: in vivo functions in neuronal cell migration and axon guidance. *Adv Exp Med Biol 515*, 13-31.

Barberis, D., Artigiani, S., Casazza, A., Corso, S., Giordano, S., Love, C. A., Jones, E. Y., Comoglio, P. M., and Tamagnone, L. (2004). Plexin signaling hampers integrin-based adhesion, leading to Rho-kinase independent cell rounding, and inhibiting lamellipodia extension and cell motility. *FASEB J 18*, 592-594.

Beech, R. D., Cleary, M. A., Treloar, H. B., Eisch, A. J., Harrist, A. V., Zhong, W., Greer, C. A., Duman, R. S., and Picciotto, M. R. (2004). Nestin promoter/enhancer directs transgene expression to precursors of adult generated periglomerular neurons. *J Comp Neurol 475*, 128-141.

Benson, D. L., and Tanaka, H. (1998). N-cadherin redistribution during synaptogenesis in hippocampal neurons. *J Neurosci 18*, 6892-6904.

Bolteus, A. J., and Bordey, A. (2004). GABA release and uptake regulate neuronal precursor migration in the postnatal subventricular zone. *J Neurosci 24*, 7623-7631.

Bonfanti, L., and Theodosis, D. T. (1994). Expression of polysialylated neural cell adhesion molecule by proliferating cells in the subependymal layer of the adult rat, in its rostral extension and in the olfactory bulb. *Neuroscience 62*, 291-305.

Bretscher, A., Chambers, D., Nguyen, R., and Reczek, D. (2000). ERM-Merlin and EBP50 protein families in plasma membrane organization and function. *Annu Rev Cell Dev Biol 16*, 113-143.

Bretscher, A., Edwards, K., and Fehon, R. G. (2002). ERM proteins and merlin: integrators at the cell cortex. *Nat Rev Mol Cell Biol 3*, 586-599.

Bretscher, A., Reczek, D., and Berryman, M. (1997). Ezrin: a protein requiring conformational activation to link microfilaments to the plasma membrane in the assembly of cell surface structures. *Journal of Cell Science 110*, 3011-3018.

Brose, K., Bland, K. S., Wang, K. H., Arnott, D., Henzel, W., Goodman, C. S., Tessier-Lavigne, M., and Kidd, T. (1999). Slit proteins bind Robo receptors and have an evolutionarily conserved role in repulsive axon guidance. *Cell 96*, 795-806.

Brunet, A., Bonni, A., Zigmond, M. J., Lin, M. Z., Juo, P., Hu, L. S., Anderson, M. J., Arden, K. C., Blenis, J., and Greenberg, M. E. (1999). Akt promotes cell survival by phosphorylating and inhibiting a Forkhead transcription factor. *Cell 96*, 857-868.

Buonanno, A., and Fischbach, G. D. (2001). Neuregulin and ErbB receptor signaling pathways in the nervous system. *Curr Opin Neurobiol 11*, 287-296.

Cameron, H. A., and Gould, E. (1994). Adult neurogenesis is regulated by adrenal steroids in the dentate gyrus. *Neuroscience 61*, 203-209.

Cantley, L. C. (2002). The phosphoinositide 3-kinase pathway. *Science 296*, 1655-1657.

Cardone, M. H., Roy, N., Stennicke, H. R., Salvesen, G. S., Franke, T. F., Stanbridge, E., Frisch, S., and Reed, J. C. (1998). Regulation of cell death protease caspase-9 by phosphorylation. *Science 282*, 1318-1321.

Carpenter, G. (2003). ErbB-4: mechanism of action and biology. *Exp Cell Res 284*, 66-77.

Caviness, V. S., Jr., and Sidman, R. L. (1972). Olfactory structures of the forebrain in the reeler mutant mouse. *J Comp Neurol 145*, 85-104.

Chauvet, N., Prieto, M., Fabre, C., Noren, N. K., and Privat, A. (2003). Distribution of p120 catenin during rat brain development: potential role in regulation of cadherin-mediated adhesion and actin cytoskeleton organization. *Mol Cell Neurosci 22*, 467-486.

Chen, H. C., and Guan, J. L. (1994). Association of focal adhesion kinase with its potential substrate phosphatidylinositol 3-kinase. *Proc Natl Acad Sci U S A 91*, 10148-10152.

Cheng, L., Itoh, K., and Lemmon, V. (2005a). L1-Mediated Branching Is Regulated by Two Ezrin-Radixin-Moesin (ERM)-Binding Sites, the RSLE Region and a Novel Juxtamembrane ERM-Binding Region. *J Neurosci 25*, 395-403.

Cheng, L. C., Tavazoie, M., and Doetsch, F. (2005b). Stem cells from epigenetics to microRNAs. *Neuron 46*, 363-367.

Cheng, M. Y., Bullock, C. M., Li, C., Lee, A. G., Bermak, J. C., Belluzzi, J., Weaver, D. R., Leslie, F. M., and Zhou, Q. Y. (2002). Prokineticin 2 transmits the behavioural circadian rhythm of the suprachiasmatic nucleus. *Nature 417*, 405-410.

Chenn, A., and Walsh, C. A. (2002). Regulation of cerebral cortical size by control of cell cycle exit in neural precursors. *Science 297*, 365-369.

Chenn, A., and Walsh, C. A. (2003). Increased neuronal production, enlarged forebrains and cytoarchitectural distortions in beta-catenin overexpressing transgenic mice. *Cereb Cortex 13*, 599-606.

Cheresh, D. A., Leng, J., and Klemke, R. L. (1999). Regulation of cell contraction and membrane ruffling by distinct signals in migratory cells. *J Cell Biol 146*, 1107-1116.

Choi, Y., Borghesani, P. R., Chan, J. A., and Segal, R. A. (2005). Migration from a Mitogenic Niche Promotes Cell-Cycle Exit. *J Neurosci 25*, 10437-10445.

Coffer, P. J., Jin, J., and Woodgett, J. R. (1998). Protein kinase B (c-Akt): a multifunctional mediator of phosphatidylinositol 3-kinase activation. *Biochem J 335*, 1-13.

Conover, J. C., Doetsch, F., Garcia-Verdugo, J. M., Gale, N. W., Yancopoulos, G. D., and Alvarez-Buylla, A. (2000). Disruption of Eph/ephrin signaling affects migration and proliferation in the adult subventricular zone. *Nat Neurosci 3*, 1091-1097.

Corotto, F. S., Henegar, J. R., and Maruniak, J. A. (1994). Odor deprivation leads to reduced neurogenesis and reduced neuronal survival in the olfactory bulb of the adult mouse. *Neuroscience 61*, 739-744.

Coskun, V., and Luskin, M. B. (2002). Intrinsic and extrinsic regulation of the proliferation and differentiation of cells in the rodent rostral migratory stream. *J Neurosci Res 69*, 795-802.

Cowan, C. W., Shao, Y. R., Sahin, M., Shamah, S. M., Lin, M. Z., Greer, P. L., Gao, S., Griffith, E. C., Brugge, J. S., and Greenberg, M. E. (2005). Vav family GEFs link activated Ephs to endocytosis and axon guidance. *Neuron 46*, 205-217.

Cremer, H., Lange, R., Christoph, A., Plomann, M., Vopper, G., Roes, J., Brown, R., Baldwin, S., Kraemer, P., Scheff, S., and et al. (1994). Inactivation of the N-CAM gene in mice results in size reduction of the olfactory bulb and deficits in spatial learning. *Nature 367*, 455-459.

Cross, D. A., Alessi, D. R., Cohen, P., Andjelkovich, M., and Hemmings, B. A. (1995). Inhibition of glycogen synthase kinase-3 by insulin mediated by protein kinase B. *Nature 378*, 785-789.

Daniel, J. M., and Reynolds, A. B. (1995). The tyrosine kinase substrate p120cas binds directly to E-cadherin but not to the adenomatous polyposis coli protein or alpha-catenin. *Mol Cell Biol 15*, 4819-4824.

Davy, A., Gale, N. W., Murray, E. W., Klinghoffer, R. A., Soriano, P., Feuerstein, C., and Robbins, S. M. (1999). Compartmentalized signaling by GPI-anchored ephrin-A5 requires the Fyn tyrosine kinase to regulate cellular adhesion. *Genes Dev 13*, 3125-3135.

Davy, A., and Robbins, S. M. (2000). Ephrin-A5 modulates cell adhesion and morphology in an integrin-dependent manner. *EMBO J 19*, 5396-5405.

de Wit, J., and Verhaagen, J. (2003). Role of semaphorins in the adult nervous system. *Prog Neurobiol 71*, 249-267.

Dedhar, S. (2000). Cell-substrate interactions and signaling through ILK. *Curr Opin Cell Biol 12*, 250-256.

del Pozo, M. A., Alderson, N. B., Kiosses, W. B., Chiang, H. H., Anderson, R. G., and Schwartz, M. A. (2004). Integrins regulate Rac targeting by internalization of membrane domains. *Science 303*, 839-842.

Del Pozo, M. A., Kiosses, W. B., Alderson, N. B., Meller, N., Hahn, K. M., and Schwartz, M. A. (2002). Integrins regulate GTP-Rac localized effector interactions through dissociation of Rho-GDI. *Nat Cell Biol 4*, 232-239.

Delcommenne, M., Tan, C., Gray, V., Rue, L., Woodgett, J., and Dedhar, S. (1998). Phosphoinositide-3-OH kinase-dependent regulation of glycogen synthase kinase 3 and protein kinase B/AKT by the integrin-linked kinase. *Proc Natl Acad Sci U S A 95*, 11211-11216.

Dickson, T. C., Mintz, C. D., Benson, D. L., and Salton, S. R. (2002). Functional binding interaction identified between the axonal CAM L1 and members of the ERM family. *J Cell Biol 157*, 1105-1112.

Doetsch, F., Caille, I., Lim, D. A., Garcia-Verdugo, J. M., and Alvarez-Buylla, A. (1999). Subventricular zone astrocytes are neural stem cells in the adult mammalian brain. *Cell 97*, 703-716.

Downward, J. (1998). Mechanisms and consequences of activation of protein kinase B/Akt. *Curr Opin Cell Biol 10*, 262-267.

Dulabon, L., Olson, E. C., Taglienti, M. G., Eisenhuth, S., McGrath, B., Walsh, C. A., Kreidberg, J. A., and Anton, E. S. (2000). Reelin binds alpha3beta1 integrin and inhibits neuronal migration. *Neuron 27*, 33-44.

Edwards, D. C., Sanders, L. C., Bokoch, G. M., and Gill, G. N. (1999). Activation of LIM-kinase by Pak1 couples Rac/Cdc42 GTPase signalling to actin cytoskeletal dynamics. *Nat Cell Biol 1*, 253-259.

Emsley, J. G., and Hagg, T. (2003). alpha6beta1 integrin directs migration of neuronal precursors in adult mouse forebrain. *Exp Neurol 183*, 273-285.

Engelkamp, D. (2002). Cloning of three mouse Unc5 genes and their expression patterns at mid-gestation. *Mech Dev 118*, 191-197.

Eriksson, P. S., Perfilieva, E., Bjork-Eriksson, T., Alborn, A. M., Nordborg, C., Peterson, D. A., and Gage, F. H. (1998). Neurogenesis in the adult human hippocampus. *Nat Med 4*, 1313-1317.

Falls, D. L. (2003). Neuregulins: functions, forms, and signaling strategies. *Exp Cell Res 284*, 14-30.

Flames, N., Long, J. E., Garratt, A. N., Fischer, T. M., Gassmann, M., Birchmeier, C., Lai, C., Rubenstein, J. L., and Marin, O. (2004). Short- and long-range attraction of cortical GABAergic interneurons by neuregulin-1. *Neuron 44*, 251-261.

Flanagan, J. G., and Vanderhaeghen, P. (1998). The ephrins and Eph receptors in neural development. *Annu Rev Neurosci* 1998 21, 309-345.

Forster, E., Tielsch, A., Saum, B., Weiss, K. H., Johanssen, C., Graus-Porta, D., Muller, U., and Frotscher, M. (2002). Reelin, Disabled 1, and beta 1 integrins are required for the formation of the radial glial scaffold in the hippocampus. *Proc Natl Acad Sci U S A 99*, 13178-13183.

Fujimoto, I., Bruses, J. L., and Rutishauser, U. (2001). Regulation of cell adhesion by polysialic acid. Effects on cadherin, immunoglobulin cell adhesion molecule, and integrin function and independence from neural cell adhesion molecule binding or signaling activity. *J Biol Chem 276*, 31745-31751.

Fujisawa, H., and Kitsukawa, T. (1998). Receptors for collapsin/semaphorins. *Curr Opin Neurobiol 8*, 587-592.

Gautreau, A., Louvard, D., and Arpin, M. (2002). ERM proteins and NF2 tumor suppressor: the Yin and Yang of cortical actin organization and cell growth signaling. *Curr Opin Cell Biol 14*, 104-109.

Gavard, J., Marthiens, V., Monnet, C., Lambert, M., and Mege, R. M. (2004). N-cadherin activation substitutes for the cell contact control in cell cycle arrest and myogenic differentiation: involvement of p120 and beta-catenin. *J Biol Chem 279*, 36795-36802.

Giachino, C., De Marchis, S., Giampietro, C., Parlato, R., Perroteau, I., Schutz, G., Fasolo, A., and Peretto, P. (2005). cAMP Response Element-Binding Protein Regulates Differentiation and Survival of Newborn Neurons in the Olfactory Bulb. *J Neurosci 25*, 10105-10118.

Giger, R. J., Pasterkamp, R. J., Heijnen, S., Holtmaat, A. J., and Verhaagen, J. (1998). Anatomical distribution of the chemorepellent semaphorin III/collapsin-1 in the adult rat and human brain: predominant expression in structures of the olfactory-hippocampal pathway and the motor system. *J Neurosci Res 52*, 27-42.

Gould, K. L., Bretscher, A., Esch, F. S., and Hunter, T. (1989). cDNA cloning and sequencing of the protein-tyrosine kinase substrate, ezrin, reveals homology to band 4.1. *EMBO J 8*, 4133-4142.

Grosheva, I., Shtutman, M., Elbaum, M., and Bershadsky, A. D. (2001). p120 catenin affects cell motility via modulation of activity of Rho-family GTPases: a link between cell-cell contact formation and regulation of cell locomotion. *J Cell Sci 114*, 695-707.

Guan, J. L. (2004). Cell biology. Integrins, rafts, Rac, and Rho. *Science 303*, 773-774.

Guan, K. L., and Rao, Y. (2003). Signalling mechanisms mediating neuronal responses to guidance cues. *Nat Rev Neurosci 4*, 941-956.

Hack, I., Bancila, M., Loulier, K., Carroll, P., and Cremer, H. (2002). Reelin is a detachment signal in tangential chain-migration during postnatal neurogenesis. *Nature Neuroscience 5*, 939-945.

Hack, M. A., Saghatelyan, A., de Chevigny, A., Pfeifer, A., Ashery-Padan, R., Lledo, P. M., and Gotz, M. (2005). Neuronal fate determinants of adult olfactory bulb neurogenesis. *Nat Neurosci. 8*, 865–872

Haegele, L., Ingold, B., Naumann, H., Tabatabai, G., Ledermann, B., and Brandner, S. (2003). Wnt signalling inhibits neural differentiation of embryonic stem cells by controlling bone morphogenetic protein expression. *Mol Cell Neurosci 24*, 696-708.

Hall, A. (1998). Rho GTPases and the actin cytoskeleton. *Science 279*, 509-514.

Hartwig, J. H., Bokoch, G. M., Carpenter, C. L., Janmey, P. A., Taylor, L. A., Toker, A., and Stossel, T. P. (1995). Thrombin receptor ligation and activated Rac uncap actin filament barbed ends through phosphoinositide synthesis in permeabilized human platelets. *Cell 82*, 643-653.

Haspel, J., and Grumet, M. (2003). The L1CAM extracellular region: a multi-domain protein with modular and cooperative binding modes. *Front Biosci 8*, s1210-1225.

He, Z., Wang, K. C., Koprivica, V., Ming, G., and Song, H. J. (2002). Knowing how to navigate: mechanisms of semaphorin signaling in the nervous system. *Sci STKE 2002*, RE1.

Hemmings, B. A. (1997). Akt signaling: linking membrane events to life and death decisions. *Science 275*, 628-630.

Hiesberger, T., Trommsdorff, M., Howell, B. W., Goffinet, A., Mumby, M. C., Cooper, J. A., and Herz, J. (1999). Direct binding of Reelin to VLDL receptor and ApoE receptor 2 induces tyrosine phosphorylation of disabled-1 and modulates tau phosphorylation. *Neuron 24*, 481-489.

Hirao, M., Sato, N., Kondo, T., Yonemura, S., Monden, M., Sasaki, T., Takai, Y., and Tsukita, S. (1996). Regulation mechanism of ERM (ezrin/radixin/moesin) protein/plasma membrane association: possible involvement of phosphatidylinositol turnover and Rho-dependent signaling pathway. *J Cell Biol 135*, 37-51.

Hiscox, S., and Jiang, W. G. (1999). Ezrin regulates cell-cell and cell-matrix adhesion, a possible role with E-cadherin/beta-catenin. *J Cell Sci 112 Pt 18*, 3081-3090.

Hu, H. (1999). Chemorepulsion of neuronal migration by Slit2 in the developing mammalian forebrain. *Neuron 23*, 703-711.

Hu, H., Tomasiewicz, H., Magnuson, T., and Rutishauser, U. (1996). The role of polysialic acid in migration of olfactory bulb interneuron precursors in the subventricular zone. *Neuron 16,* 735-743.

Huai, J., and Drescher, U. (2001). An ephrin-A-dependent signaling pathway controls integrin function and is linked to the tyrosine phosphorylation of a 120-kDa protein. *J Biol Chem 276*, 6689-6694.

Hynes, R. O. (2002). Integrins: bidirectional, allosteric signaling machines. Cell 110, 673-687.

Irie, F., and Yamaguchi, Y. (2002). EphB receptors regulate dendritic spine development via intersectin, Cdc42 and N-WASP. *Nat Neurosci 5*, 1117-1118.

Itoh, A., Miyabayashi, T., Ohno, M., and Sakano, S. (1998). Cloning and expressions of three mammalian homologues of Drosophila slit suggest possible roles for Slit in the formation and maintenance of the nervous system. *Brain Res Mol Brain Res 62*, 175-186.

Jacques, T. S., Relvas, J. B., Nishimura, S., Pytela, R., Edwards, G. M., Streuli, C. H., and ffrench-Constant, C. (1998). Neural precursor cell chain migration and division are regulated through different beta1 integrins. *Development 125*, 3167-3177.

Jenkins, S. M., Kizhatil, K., Kramarcy, N. R., Sen, A., Sealock, R., and Bennett, V. (2001). FIGQY phosphorylation defines discrete populations of L1 cell adhesion molecules at sites of cell-cell contact and in migrating neurons. *J Cell Sci 114*, 3823-3835.

Jin, Z., and Strittmatter, S. M. (1997). Rac1 mediates collapsin-1-induced growth cone collapse. *J Neurosci 17*, 6256-6263.

Johnson, M. W., Miyata, H., and Vinters, H. V. (2002). Ezrin and moesin expression within the developing human cerebrum and tuberous sclerosis-associated cortical tubers. *Acta Neuropathol (Berl) 104*, 188-196.

Jones, F. S., and Jones, P. L. (2000). The tenascin family of ECM glycoproteins: structure, function, and regulation during embryonic development and tissue remodeling. *Dev Dyn 218*, 235-259.

Jurney, W. M., Gallo, G., Letourneau, P. C., and McLoon, S. C. (2002). Rac1-mediated endocytosis during ephrin-A2- and semaphorin 3A-induced growth cone collapse. *J Neurosci 22*, 6019-6028.

Kaibuchi, K., Kuroda, S., and Amano, M. (1999). Regulation of the cytoskeleton and cell adhesion by the Rho family GTPases in mammalian cells. *Annu Rev Biochem 68*, 459-486.

Kaverina, I., Krylyshkina, O., and Small, J. V. (2002). Regulation of substrate adhesion dynamics during cell motility. *Int J Biochem Cell Biol 34*, 746-761.

Keino-Masu, K., Masu, M., Hinck, L., Leonardo, E. D., Chan, S. S., Culotti, J. G., and Tessier-Lavigne, M. (1996). Deleted in Colorectal Cancer (DCC) encodes a netrin receptor. *Cell 87*, 175-185.

Kenwrick, S., Watkins, A., and De Angelis, E. (2000). Neural cell recognition molecule L1: relating biological complexity to human disease mutations. *Hum Mol Genet 9*, 879-886.

Klemke, R. L., Cai, S., Giannini, A. L., Gallagher, P. J., de Lanerolle, P., and Cheresh, D. A. (1997). Regulation of cell motility by mitogen-activated protein kinase. *J Cell Biol 137*, 481-492.

Kohwi, M., Osumi, N., Rubenstein, J. L. R., and Alvarez-Buylla, A. (2005). Pax6 Is Required for Making Specific Subpopulations of Granule and Periglomerular Neurons in the Olfactory Bulb. *J Neurosci 25*, 6997-7003.

Kolodkin, A. L., Matthes, D. J., and Goodman, C. S. (1993). The semaphorin genes encode a family of transmembrane and secreted growth cone guidance molecules. *Cell 75*, 1389-1399.

Kornack, D. R., and Rakic, P. (1999). Continuation of neurogenesis in the hippocampus of the adult macaque monkey. *Proc Natl Acad Sci U S A 96*, 5768-5773.

Kuhn, H. G., Dickinson-Anson, H., and Gage, F. H. (1996). Neurogenesis in the dentate gyrus of the adult rat: age-related decrease of neuronal progenitor proliferation. *J Neurosci 16*, 2027-2033.

Kuhn, T. B., Brown, M. D., Wilcox, C. L., Raper, J. A., and Bamburg, J. R. (1999). Myelin and collapsin-1 induce motor neuron growth cone collapse through different pathways: inhibition of collapse by opposing mutants of rac1. *J Neurosci 19*, 1965-1975.

Lambert de Rouvroit, C., and Goffinet, A. M. (1998). A new view of early cortical development. *Biochem Pharmacol 56*, 1403-1409.

Larson, J., Hoffman, J. S., Guidotti, A., and Costa, E. (2003). Olfactory discrimination learning deficit in heterozygous reeler mice. *Brain Res 971*, 40-46.

Lauffenburger, D. A., and Horwitz, A. F. (1996). Cell migration: a physically integrated molecular process. *Cell 84*, 359-369.

LeCouter, J., Lin, R., Tejada, M., Frantz, G., Peale, F., Hillan, K. J., and Ferrara, N. (2003). The endocrine-gland-derived VEGF homologue Bv8 promotes angiogenesis in the testis: Localization of Bv8 receptors to endothelial cells. *Proc Natl Acad Sci U S A 100*, 2685-2690.

Lemasson, M., Saghatelyan, A., Olivo-Marin, J.-C., and Lledo, P.-M. (2005). Neonatal and Adult Neurogenesis Provide Two Distinct Populations of Newborn Neurons to the Mouse Olfactory Bulb. *J Neurosci 25*, 6816-6825.

Leonardo, E. D., Hinck, L., Masu, M., Keino-Masu, K., Ackerman, S. L., and Tessier-Lavigne, M. (1997). Vertebrate homologues of C. elegans UNC-5 are candidate netrin receptors. *Nature 386*, 833-838.

Li, M., Bullock, C. M., Knauer, D. J., Ehlert, F. J., and Zhou, Q. Y. (2001). Identification of two prokineticin cDNAs: recombinant proteins potently contract gastrointestinal smooth muscle. *Mol Pharmacol 59*, 692-698.

Li, S., Guan, J. L., and Chien, S. (2005). Biochemistry and biomechanics of cell motility. *Annu Rev Biomed Eng 7*, 105-150.

Li, W., Lee, J., Vikis, H. G., Lee, S. H., Liu, G., Aurandt, J., Shen, T. L., Fearon, E. R., Guan, J. L., Han, M., et al. (2004). Activation of FAK and Src are receptor-proximal events required for netrin signaling. *Nat Neurosci 7*, 1213-1221.

Li, X., Saint-Cyr-Proulx, E., Aktories, K., and Lamarche-Vane, N. (2002). Rac1 and Cdc42 but not RhoA or Rho kinase activities are required for neurite outgrowth induced by the Netrin-1 receptor DCC (deleted in colorectal cancer) in N1E-115 neuroblastoma cells. *J Biol Chem 277*, 15207-15214.

Liang, J., Zubovitz, J., Petrocelli, T., Kotchetkov, R., Connor, M. K., Han, K., Lee, J. H., Ciarallo, S., Catzavelos, C., Beniston, R., et al. (2002). PKB/Akt phosphorylates p27, impairs nuclear import of p27 and opposes p27-mediated G1 arrest. *Nat Med 8*, 1153-1160.

Lim, D. A., and Alvarez-Buylla, A. (1999). Interaction between astrocytes and adult subventricular zone precursors stimulates neurogenesis. *Proc Natl Acad Sci U S A 96*, 7526-7531.

Lin, D. C., Bullock, C. M., Ehlert, F. J., Chen, J. L., Tian, H., and Zhou, Q. Y. (2002). Identification and molecular characterization of two closely related G protein-coupled receptors activated by prokineticins/endocrine gland vascular endothelial growth factor. *J Biol Chem 277*, 19276-19280.

Liu, G., Beggs, H., Jurgensen, C., Park, H. T., Tang, H., Gorski, J., Jones, K. R., Reichardt, L. F., Wu, J., and Rao, Y. (2004). Netrin requires focal adhesion kinase and Src family kinases for axon outgrowth and attraction. *Nat Neurosci 7*, 1222-1232.

Liu, G., and Rao, Y. (2003). Neuronal migration from the forebrain to the olfactory bulb requires a new attractant persistent in the olfactory bulb. *J Neurosci 23*, 6651-6659.

Liu, X., Wang, Q., Haydar, T. F., and Bordey, A. (2005). Nonsynaptic GABA signaling in postnatal subventricular zone controls proliferation of GFAP-expressing progenitors. *Nat Neurosci 8*, 1179-1187.

Lledo, P. M., and Saghatelyan, A. (2005). Integrating new neurons into the adult olfactory bulb: joining the network, life-death decisions, and the effects of sensory experience. *Trends Neurosci 28*, 248-254.

Lodish , H., Darnell, J., Matsudaira, P., Berk, A., and Zipursky, S. (2003). The Dynamics of Actin Assembly. In *Molecular cell biology* (New York, W.H. Freeman and Company), pp. 784-791.

Logan, C. Y., and Nusse, R. (2004). The Wnt signaling pathway in development and disease. *Annu Rev Cell Dev Biol 20*, 781-810.

Lois, C., and Alvarez-Buylla, A. (1994). Long-distance neuronal migration in the adult mammalian brain. *Science 264*, 1145-1148.

Lois, C., Garcia-Verdugo, J. M., and Alvarez-Buylla, A. (1996). Chain migration of neuronal precursors. *Science 271*, 978-981.

Long, H., Sabatier, C., Ma, L., Plump, A., Yuan, W., Ornitz, D. M., Tamada, A., Murakami, F., Goodman, C. S., and Tessier-Lavigne, M. (2004). Conserved roles for Slit and Robo proteins in midline commissural axon guidance. *Neuron 42,* 213-223.

Louvet-Vallee, S. (2000). ERM proteins: from cellular architecture to cell signaling. *Biol Cell 92*, 305-316.

Luskin, M. B. (1993). Restricted proliferation and migration of postnatally generated neurons derived from the forebrain subventricular zone. *Neuron 11*, 173-189.

Maeda, M., Matsui, T., Imamura, M., and Tsukita, S. (1999). Expression level, subcellular distribution and rho-GDI binding affinity of merlin in comparison with Ezrin/Radixin/Moesin proteins. *Oncogene 18*, 4788-4797.

Maekawa, M., Ishizaki, T., Boku, S., Watanabe, N., Fujita, A., Iwamatsu, A., Obinata, T., Ohashi, K., Mizuno, K., and Narumiya, S. (1999). Signaling from Rho to the actin cytoskeleton through protein kinases ROCK and LIM-kinase. *Science 285*, 895-898.

Marillat, V., Cases, O., Nguyen-Ba-Charvet, K. T., Tessier-Lavigne, M., Sotelo, C., and Chedotal, A. (2002). Spatiotemporal expression patterns of slit and robo genes in the rat brain. *J Comp Neurol 442*, 130-155.

Marin, O., and Rubenstein, J. L. (2001). A long, remarkable journey: tangential migration in the telencephalon. *Nat Rev Neurosci 2*, 780-790.

Marin, O., and Rubenstein, J. L. (2003). Cell migration in the forebrain. *Annu Rev Neurosci 26*, 441-483.

Marin, O., Yaron, A., Bagri, A., Tessier-Lavigne, M., and Rubenstein, J. L. (2001). Sorting of striatal and cortical interneurons regulated by semaphorin-neuropilin interactions. *Science 293*, 872-875.

Mason, H. A., Ito, S., and Corfas, G. (2001). Extracellular signals that regulate the tangential migration of olfactory bulb neuronal precursors: inducers, inhibitors, and repellents. *J Neurosci 21*, 7654-7663.

Matarredona, E. R., Murillo-Carretero, M., Moreno-Lopez, B., and Estrada, C. (2005). Role of nitric oxide in subventricular zone neurogenesis. *Brain Res Brain Res Rev 49*, 355-366.

Menezes, J. R., Smith, C. M., Nelson, K. C., and Luskin, M. B. (1995). The division of neuronal progenitor cells during migration in the neonatal mammalian forebrain. *Mol Cell Neurosci 6*, 496-508.

Miki, H., Yamaguchi, H., Suetsugu, S., and Takenawa, T. (2000). IRSp53 is an essential intermediate between Rac and WAVE in the regulation of membrane ruffling. *Nature 408*, 732-735.

Mintz, C. D., Dickson, T. C., Gripp, M. L., Salton, S. R., and Benson, D. L. (2003). ERMs colocalize transiently with L1 during neocortical axon outgrowth. *J Comp Neurol 464*, 438-448.

Miranti, C. K., and Brugge, J. S. (2002). Sensing the environment: a historical perspective on integrin signal transduction. *Nat Cell Biol 4*, E83-90.

Mitchison, T. J., and Cramer, L. P. (1996). Actin-based cell motility and cell locomotion. *Cell 84*, 371-379.

Murai, K. K., and Pasquale, E. B. (2005). New Exchanges in Eph-Dependent Growth Cone Dynamics. *Neuron 46*, 161-163.

Murase, S., and Horwitz, A. F. (2002). Deleted in colorectal carcinoma and differentially expressed integrins mediate the directional migration of neural precursors in the rostral migratory stream. *J Neurosci 22*, 3568-3579.

Negishi, M., Oinuma, I., and Katoh, H. (2005). Plexins: axon guidance and signal transduction. *Cell Mol Life Sci 62*, 1363-1371.

Nelson, W. J., and Nusse, R. (2004). Convergence of Wnt, {beta}-Catenin, and Cadherin Pathways. *Science 303*, 1483-1487.

Ng, K. L., Li, J. D., Cheng, M. Y., Leslie, F. M., Lee, A. G., and Zhou, Q. Y. (2005). Dependence of olfactory bulb neurogenesis on prokineticin 2 signaling. *Science 308*, 1923-1927.

Nguyen, D. H., Catling, A. D., Webb, D. J., Sankovic, M., Walker, L. A., Somlyo, A. V., Weber, M. J., and Gonias, S. L. (1999). Myosin light chain kinase functions downstream of Ras/ERK to promote migration of urokinase-type plasminogen activator-stimulated cells in an integrin-selective manner. *J Cell Biol 146*, 149-164.

Nguyen-Ba-Charvet, K. T., Picard-Riera, N., Tessier-Lavigne, M., Baron-Van Evercooren, A., Sotelo, C., and Chedotal, A. (2004). Multiple roles for slits in the control of cell migration in the rostral migratory stream. *J Neurosci 24*, 1497-1506.

Ni, C. Y., Murphy, M. P., Golde, T. E., and Carpenter, G. (2001). gamma -Secretase cleavage and nuclear localization of ErbB-4 receptor tyrosine kinase. *Science 294*, 2179-2181.

Noren, N. K., Liu, B. P., Burridge, K., and Kreft, B. (2000). p120 catenin regulates the actin cytoskeleton via Rho family GTPases. *J Cell Biol 150*, 567-580.

Noren, N. K., and Pasquale, E. B. (2004). Eph receptor-ephrin bidirectional signals that target Ras and Rho proteins. *Cell Signal 16*, 655-666.

O'Leary, D. D., and Wilkinson, D. G. (1999). Eph receptors and ephrins in neural development. *Curr Opin Neurobiol 9*, 65-73.

O'Rourke, N. A. (1996). Neuronal chain gangs: homotypic contacts support migration into the olfactory bulb. *Neuron 16*, 1061-1064.

Ono, K., Tomasiewicz, H., Magnuson, T., and Rutishauser, U. (1994). N-CAM mutation inhibits tangential neuronal migration and is phenocopied by enzymatic removal of polysialic acid. *Neuron 13,* 595-609.

Palazzo, A. F., Cook, T. A., Alberts, A. S., and Gundersen, G. G. (2001). mDia mediates Rho-regulated formation and orientation of stable microtubules. *Nat Cell Biol 3,* 723-729.

Palma, V., Lim, D. A., Dahmane, N., Sanchez, P., Brionne, T. C., Herzberg, C. D., Gitton, Y., Carleton, A., Alvarez-Buylla, A., and Ruiz i Altaba, A. (2005). Sonic hedgehog controls stem cell behavior in the postnatal and adult brain. *Development 132,* 335-344.

Palmer, A., and Klein, R. (2003). Multiple roles of ephrins in morphogenesis, neuronal networking, and brain function. *Genes Dev 17,* 1429-1450.

Palmer, A., Zimmer, M., Erdmann, K. S., Eulenburg, V., Porthin, A., Heumann, R., Deutsch, U., and Klein, R. (2002). EphrinB phosphorylation and reverse signaling: regulation by Src kinases and PTP-BL phosphatase. *Mol Cell 9,* 725-737.

Palmer, T. D., Willhoite, A. R., and Gage, F. H. (2000). Vascular niche for adult hippocampal neurogenesis. *J Comp Neurol 425,* 479-494.

Pappas, G. D., Kriho, V., Liu, W. S., Tremolizzo, L., Lugli, G., and Larson, J. (2003). Immunocytochemical localization of reelin in the olfactory bulb of the heterozygous reeler mouse: an animal model for schizophrenia. *Neurol Res 25,* 819-830.

Park, H. T., Wu, J., and Rao, Y. (2002). Molecular control of neuronal migration. *Bioessays 24,* 821-827.

Parrini, M. C., Lei, M., Harrison, S. C., and Mayer, B. J. (2002). Pak1 kinase homodimers are autoinhibited in trans and dissociated upon activation by Cdc42 and Rac1. *Mol Cell 9,* 73-83.

Pasterkamp, R. J., and Kolodkin, A. L. (2003). Semaphorin junction: making tracks toward neural connectivity. *Curr Opin Neurobiol 13,* 79-89.

Penzes, P., Beeser, A., Chernoff, J., Schiller, M. R., Eipper, B. A., Mains, R. E., and Huganir, R. L. (2003). Rapid induction of dendritic spine morphogenesis by trans-synaptic ephrinB-EphB receptor activation of the Rho-GEF kalirin. *Neuron 37,* 263-274.

Peretto, P., Merighi, A., Fasolo, A., and Bonfanti, L. (1997). Glial tubes in the rostral migratory stream of the adult rat. *Brain Res Bull 42,* 9-21.

Perrot, V., Vazquez-Prado, J., and Gutkind, J. S. (2002). Plexin B regulates Rho through the guanine nucleotide exchange factors leukemia-associated Rho GEF (LARG) and PDZ-RhoGEF. *J Biol Chem 277,* 43115-43120.

Pesheva, P., and Probstmeier, R. (2000). The yin and yang of tenascin-R in CNS development and pathology. *Prog Neurobiol 61,* 465-493.

Petridis, A. K., El-Maarouf, A., and Rutishauser, U. (2004). Polysialic acid regulates cell contact-dependent neuronal differentiation of progenitor cells from the subventricular zone. *Dev Dyn 230,* 675-684.

Pollard, T. D., Blanchoin, L., and Mullins, R. D. (2000). Molecular mechanisms controlling actin filament dynamics in nonmuscle cells. Annu Rev Biophys Biomol Struct 29, 545-576.

Pollard, T. D., and Borisy, G. G. (2003). Cellular motility driven by assembly and disassembly of actin filaments. *Cell 112,* 453-465.

Ponti, A., Machacek, M., Gupton, S. L., Waterman-Storer, C. M., and Danuser, G. (2004). Two Distinct Actin Networks Drive the Protrusion of Migrating Cells. *Science 305*, 1782-1786.

Porfiri, E., Rubinfeld, B., Albert, I., Hovanes, K., Waterman, M., and Polakis, P. (1997). Induction of a beta-catenin-LEF-1 complex by wnt-1 and transforming mutants of beta-catenin. *Oncogene 15*, 2833-2839.

Przyborski, S. A., Knowles, B. B., and Ackerman, S. L. (1998). Embryonic phenotype of Unc5h3 mutant mice suggests chemorepulsion during the formation of the rostral cerebellar boundary. *Development 125*, 41-50.

Rafelski, S. M., and Theriot, J. A. (2004). Crawling toward a unified model of cell mobility: spatial and temporal regulation of actin dynamics. *Annu Rev Biochem 73*, 209-239.

Raftopoulou, M., and Hall, A. (2004). Cell migration: Rho GTPases lead the way. *Dev Biol 265*, 23-32.

Rakic, P., and Caviness, V. S., Jr. (1995). Cortical development: view from neurological mutants two decades later. *Neuron 14*, 1101-1104.

Redies, C. (2000). Cadherins in the central nervous system. *Prog Neurobiol 61*, 611-648.

Ren, X. R., Ming, G. L., Xie, Y., Hong, Y., Sun, D. M., Zhao, Z. Q., Feng, Z., Wang, Q., Shim, S., Chen, Z. F., et al. (2004). Focal adhesion kinase in netrin-1 signaling. *Nat Neurosci 7*, 1204-1212.

Reynolds, A. B., Daniel, J., McCrea, P. D., Wheelock, M. J., Wu, J., and Zhang, Z. (1994). Identification of a new catenin: the tyrosine kinase substrate p120cas associates with E-cadherin complexes. *Mol Cell Biol 14*, 8333-8342.

Rhee, J., Mahfooz, N. S., Arregui, C., Lilien, J., Balsamo, J., and VanBerkum, M. F. (2002). Activation of the repulsive receptor Roundabout inhibits N-cadherin-mediated cell adhesion. *Nat Cell Biol 4*, 798-805.

Rice, D. S., and Curran, T. (2001). Role of the reelin signaling pathway in central nervous system development. *Annu Rev Neurosci 24*, 1005-1039.

Ridley, A. J. (2001). Rho GTPases and cell migration. *J Cell Sci 114*, 2713-2722.

Ridley, A. J., Schwartz, M. A., Burridge, K., Firtel, R. A., Ginsberg, M. H., Borisy, G., Parsons, J. T., and Horwitz, A. R. (2003). Cell migration: integrating signals from front to back. *Science 302*, 1704-1709.

Riese, D. J., 2nd, and Stern, D. F. (1998). Specificity within the EGF family/ErbB receptor family signaling network. *Bioessays 20*, 41-48.

Rio, C., Buxbaum, J. D., Peschon, J. J., and Corfas, G. (2000). Tumor necrosis factor-alpha-converting enzyme is required for cleavage of erbB4/HER4. *J Biol Chem 275*, 10379-10387.

Rousselot, P., Lois, C., and Alvarez-Buylla, A. (1995). Embryonic (PSA) N-CAM reveals chains of migrating neuroblasts between the lateral ventricle and the olfactory bulb of adult mice. *J Comp Neurol 351*, 51-61.

Rugarli, E. I., and Ballabio, A. (1995). Reelin: a novel extracellular matrix protein involved in brain lamination. *Bioessays 17*, 832-834.

Saghatelyan, A., de Chevigny, A., Schachner, M., and Lledo, P. M. (2004). Tenascin-R mediates activity-dependent recruitment of neuroblasts in the adult mouse forebrain. *Nat Neurosci 7*, 347-356.

Sahin, M., Greer, P. L., Lin, M. Z., Poucher, H., Eberhart, J., Schmidt, S., Wright, T. M., Shamah, S. M., O'Connell, S., Cowan, C. W., et al. (2005). Eph-dependent tyrosine phosphorylation of ephexin1 modulates growth cone collapse. *Neuron 46*, 191-204.

Sanders, L. C., Matsumura, F., Bokoch, G. M., and de Lanerolle, P. (1999). Inhibition of myosin light chain kinase by p21-activated kinase. *Science 283*, 2083-2085.

Schaller, M. D. (2001). Biochemical signals and biological responses elicited by the focal adhesion kinase. *Biochim Biophys Acta 1540*, 1-21.

Schmid, R. S., and Anton, E. S. (2003). Role of integrins in the development of the cerebral cortex. *Cereb Cortex 13*, 219-224.

Schmid, R. S., Jo, R., Shelton, S., Kreidberg, J. A., and Anton, E. S. (2005). Reelin, Integrin and Dab1 Interactions during Embryonic Cerebral Cortical Development. *Cereb Cortex 15*, 1632-1636.

Schmidt, A., and Hall, A. (2002). Guanine nucleotide exchange factors for Rho GTPases: turning on the switch. *Genes Dev 16*, 1587-1609.

Serafini, T., Colamarino, S. A., Leonardo, E. D., Wang, H., Beddington, R., Skarnes, W. C., and Tessier-Lavigne, M. (1996). Netrin-1 is required for commissural axon guidance in the developing vertebrate nervous system. *Cell 87*, 1001-1014.

Serini, G., Valdembri, D., Zanivan, S., Morterra, G., Burkhardt, C., Caccavari, F., Zammataro, L., Primo, L., Tamagnone, L., Logan, M., et al. (2003). Class 3 semaphorins control vascular morphogenesis by inhibiting integrin function. *Nature 424*, 391-397.

Shamah, S. M., Lin, M. Z., Goldberg, J. L., Estrach, S., Sahin, M., Hu, L., Bazalakova, M., Neve, R. L., Corfas, G., Debant, A., and Greenberg, M. E. (2001). EphA receptors regulate growth cone dynamics through the novel guanine nucleotide exchange factor ephexin. *Cell 105*, 233-244.

Sheetz, M. P., Felsenfeld, D., Galbraith, C. G., and Choquet, D. (1999). Cell migration as a five-step cycle. *Biochem Soc Symp 65*, 233-243.

Shekarabi, M., and Kennedy, T. E. (2002). The netrin-1 receptor DCC promotes filopodia formation and cell spreading by activating Cdc42 and Rac1. *Mol Cell Neurosci 19*, 1-17.

Sheldon, M., Rice, D. S., D'Arcangelo, G., Yoneshima, H., Nakajima, K., Mikoshiba, K., Howell, B. W., Cooper, J. A., Goldowitz, D., and Curran, T. (1997). Scrambler and yotari disrupt the disabled gene and produce a reeler-like phenotype in mice. *Nature 389*, 730-733.

Shen, Q., Goderie, S. K., Jin, L., Karanth, N., Sun, Y., Abramova, N., Vincent, P., Pumiglia, K., and Temple, S. (2004). Endothelial cells stimulate self-renewal and expand neurogenesis of neural stem cells. *Science 304*, 1338-1340.

Shen, T. L., and Guan, J. L. (2001). Differential regulation of cell migration and cell cycle progression by FAK complexes with Src, PI3K, Grb7 and Grb2 in focal contacts. *FEBS Lett 499, 176-181.*

Shepard, G., and Greer, C. (1998). Olfactory Bulb. In *The synaptic organization of the brain*, G. Shepard, ed. (New York, Oxford University Press), pp. 159-204.

Shibamoto, S., Hayakawa, M., Takeuchi, K., Hori, T., Miyazawa, K., Kitamura, N., Johnson, K. R., Wheelock, M. J., Matsuyoshi, N., Takeichi, M., and et al. (1995). Association of p120, a tyrosine kinase substrate, with E-cadherin/catenin complexes. *J Cell Biol 128*, 949-957.

Small, J. V., and Kaverina, I. (2005). Polarized Cell Motility: Microtubules show the way. In *Cell migration in development and disease*. D. Wedlich, ed. (Weinheim, Wiley-VCH), pp. 15-31.

Soga, T., Matsumoto, S., Oda, T., Saito, T., Hiyama, H., Takasaki, J., Kamohara, M., Ohishi, T., Matsushime, H., and Furuichi, K. (2002). Molecular cloning and characterization of prokineticin receptors. *Biochim Biophys Acta 1579*, 173-179.

Song, H., Stevens, C. F., and Gage, F. H. (2002). Astroglia induce neurogenesis from adult neural stem cells. *Nature 417*, 39-44.

Spradling, A., Drummond-Barbosa, D., and Kai, T. (2001). Stem cells find their niche. *Nature 414*, 98-104.

Staddon, J. M., Smales, C., Schulze, C., Esch, F. S., and Rubin, L. L. (1995). p120, a p120-related protein (p100), and the cadherin/catenin complex. *J Cell Biol 130*, 369-381.

Stefansson, H., Sarginson, J., Kong, A., Yates, P., Steinthorsdottir, V., Gudfinnsson, E., Gunnarsdottir, S., Walker, N., Petursson, H., Crombie, C., et al. (2003). Association of neuregulin 1 with schizophrenia confirmed in a Scottish population. *Am J Hum Genet 72*, 83-87.

Stefansson, H., Sigurdsson, E., Steinthorsdottir, V., Bjornsdottir, S., Sigmundsson, T., Ghosh, S., Brynjolfsson, J., Gunnarsdottir, S., Ivarsson, O., Chou, T. T., et al. (2002). Neuregulin 1 and susceptibility to schizophrenia. *Am J Hum Genet 71*, 877-892.

Stein, E., and Tessier-Lavigne, M. (2001). Hierarchical organization of guidance receptors: silencing of netrin attraction by slit through a Robo/DCC receptor complex. *Science 291*, 1928-1938.

Stull, J. T., Tansey, M. G., Tang, D. C., Word, R. A., and Kamm, K. E. (1993). Phosphorylation of myosin light chain kinase: a cellular mechanism for Ca2+ desensitization. *Mol Cell Biochem* 127-128, 229-237.

Swiercz, J. M., Kuner, R., Behrens, J., and Offermanns, S. (2002). Plexin-B1 directly interacts with PDZ-RhoGEF/LARG to regulate RhoA and growth cone morphology. *Neuron 35*, 51-63.

Symons, M., and Settleman, J. (2000). Rho family GTPases: more than simple switches. *Trends Cell Biol 10*, 415-419.

Takahashi, K., Sasaki, T., Mammoto, A., Takaishi, K., Kameyama, T., Tsukita, S., and Takai, Y. (1997). Direct interaction of the Rho GDP dissociation inhibitor with ezrin/radixin/moesin initiates the activation of the Rho small G protein. *J Biol Chem 272*, 23371-23375.

Takahashi, T., Fournier, A., Nakamura, F., Wang, L. H., Murakami, Y., Kalb, R. G., Fujisawa, H., and Strittmatter, S. M. (1999). Plexin-neuropilin-1 complexes form functional semaphorin-3A receptors. *Cell 99*, 59-69.

Takeuchi, K., Kawashima, A., Nagafuchi, A., and Tsukita, S. (1994). Structural diversity of band 4.1 superfamily members. *J Cell Sci 107*, 1921-1928.

Tamagnone, L., and Comoglio, P. M. (2000). Signalling by semaphorin receptors: cell guidance and beyond. *Trends Cell Biol 10*, 377-383.

Tamagnone, L., and Comoglio, P. M. (2004). To move or not to move? Semaphorin signalling in cell migration. *EMBO Rep 5*, 356-361.

Tamamaki, N., Fujimori, K., Nojyo, Y., Kaneko, T., and Takauji, R. (2003). Evidence that Sema3A and Sema3F regulate the migration of GABAergic neurons in the developing neocortex. *J Comp Neurol 455*, 238-248.

Tanaka, M., Ohashi, R., Nakamura, R., Shinmura, K., Kamo, T., Sakai, R., and Sugimura, H. (2004). Tiam1 mediates neurite outgrowth induced by ephrin-B1 and EphA2. *EMBO J 23*, 1075-1088.

Tate, M. C., Garcia, A. J., Keselowsky, B. G., Schumm, M. A., Archer, D. R., and LaPlaca, M. C. (2004). Specific beta1 integrins mediate adhesion, migration, and differentiation of neural progenitors derived from the embryonic striatum. *Mol Cell Neurosci 27*, 22-31.

Tessier-Lavigne, M., and Goodman, C. S. (1996). The molecular biology of axon guidance. *Science 274*, 1123-1133.

Thelen, K., Kedar, V., Panicker, A. K., Schmid, R. S., Midkiff, B. R., and Maness, P. F. (2002). The neural cell adhesion molecule L1 potentiates integrin-dependent cell migration to extracellular matrix proteins. *J Neurosci 22*, 4918-4931.

Tomasiewicz, H., Ono, K., Yee, D., Thompson, C., Goridis, C., Rutishauser, U., and Magnuson, T. (1993). Genetic deletion of a neural cell adhesion molecule variant (N-CAM-180) produces distinct defects in the central nervous system. *Neuron 11*, 1163-1174.

Tonchev, A. B., Yamashima, T., Sawamoto, K., and Okano, H. (2005). Enhanced proliferation of progenitor cells in the subventricular zone and limited neuronal production in the striatum and neocortex of adult macaque monkeys after global cerebral ischemia. *J Neurosci Res.*

Vanhaesebroeck, B., and Alessi, D. R. (2000). The PI3K-PDK1 connection: more than just a road to PKB. *Biochem J 346*, 561-576.

Viglietto, G., Motti, M. L., Bruni, P., Melillo, R. M., D'Alessio, A., Califano, D., Vinci, F., Chiappetta, G., Tsichlis, P., Bellacosa, A., et al. (2002). Cytoplasmic relocalization and inhibition of the cyclin-dependent kinase inhibitor p27(Kip1) by PKB/Akt-mediated phosphorylation in breast cancer. *Nat Med 8,* 1136-1144.

Wahl, S., Barth, H., Ciossek, T., Aktories, K., and Mueller, B. K. (2000). Ephrin-A5 induces collapse of growth cones by activating Rho and Rho kinase. *J Cell Biol 149*, 263-270.

Wang, D. D., Krueger, D. D., and Bordey, A. (2003). GABA depolarizes neuronal progenitors of the postnatal subventricular zone via GABAA receptor activation. *J Physiol 550*, 785-800.

Wang, T. W., Zhang, H., and Parent, J. M. (2005). Retinoic acid regulates postnatal neurogenesis in the murine subventricular zone-olfactory bulb pathway. *Development.*

Ware, M. L., Fox, J. W., Gonzalez, J. L., Davis, N. M., Lambert de Rouvroit, C., Russo, C. J., Chua, S. C., Jr., Goffinet, A. M., and Walsh, C. A. (1997). Aberrant splicing of a mouse disabled homolog, mdab1, in the scrambler mouse. *Neuron 19*, 239-249.

Watanabe, T., Noritake, J., and Kaibuchi, K. (2005). Regulation of microtubules in cell migration. *Trends Cell Biol 15*, 76-83.

Waterman-Storer, C. M., and Salmon, E. (1999). Positive feedback interactions between microtubule and actin dynamics during cell motility. *Curr Opin Cell Biol 11*, 61-67.

Welch, M. D., and Mullins, R. D. (2002). Cellular control of actin nucleation. *Annu Rev Cell Dev Biol 18*, 247-288.

Wichterle, H., Garcia-Verdugo, J. M., and Alvarez-Buylla, A. (1997). Direct evidence for homotypic, glia-independent neuronal migration. *Neuron 18,* 779-791.

Wiesner, S., Legate, K. R., and Fassler, R. (2005). Integrin-actin interactions. *Cell Mol Life Sci 62*, 1081-1099.

Wilkinson, D. G. (2001). Multiple roles of EPH receptors and ephrins in neural development. *Nat Rev Neurosci 2*, 155-164.

Winner, B., Cooper-Kuhn, C. M., Aigner, R., Winkler, J., and Kuhn, H. G. (2002). Long-term survival and cell death of newly generated neurons in the adult rat olfactory bulb. *Eur J Neurosci 16,* 1681-1689.

Wolf, N. S., and Trentin, J. J. (1968). Hemopoietic colony studies. V. Effect of hemopoietic organ stroma on differentiation of pluripotent stem cells. *J Exp Med 127*, 205-214.

Wong, K., Park, H. T., Wu, J. Y., and Rao, Y. (2002). Slit proteins: molecular guidance cues for cells ranging from neurons to leukocytes. *Curr Opin Genet Dev 12*, 583-591.

Wong, K., Ren, X. R., Huang, Y. Z., Xie, Y., Liu, G., Saito, H., Tang, H., Wen, L., Brady-Kalnay, S. M., Mei, L., et al. (2001). Signal transduction in neuronal migration: roles of GTPase activating proteins and the small GTPase Cdc42 in the Slit-Robo pathway. *Cell 107*, 209-221.

Wu, W., Wong, K., Chen, J., Jiang, Z., Dupuis, S., Wu, J. Y., and Rao, Y. (1999). Directional guidance of neuronal migration in the olfactory system by the protein Slit. *Nature 400*, 331-336.

Wurmser, A. E., Palmer, T. D., and Gage, F. H. (2004). Cellular interactions in the stem cell niche. *Science 304*, 1253-1255.

Yang, S. Z., Kohno, N., Yokoyama, A., Kondo, K., Hamada, H., and Hiwada, K. (2001). Decreased E-cadherin augments beta-catenin nuclear localization: studies in breast cancer cell lines. *Int J Oncol 18*, 541-548.

Yarden, Y., and Sliwkowski, M. X. (2001). Untangling the ErbB signalling network. *Nat Rev Mol Cell Biol 2*, 127-137.

Yau, H. J., Wang, H. F., Lai, C., and Liu, F. C. (2003). Neural development of the neuregulin receptor ErbB4 in the cerebral cortex and the hippocampus: preferential expression by interneurons tangentially migrating from the ganglionic eminences. *Cereb Cortex 13*, 252-264.

Yu, T. W., and Bargmann, C. I. (2001). Dynamic regulation of axon guidance. *Nat Neurosci 4*, 1169-1176.

Zamir, E., and Geiger, B. (2001). Molecular complexity and dynamics of cell-matrix adhesions. *J Cell Sci 114*, 3583-3590.

Zhao, X., Shi, Y., Tang, J., Tang, R., Yu, L., Gu, N., Feng, G., Zhu, S., Liu, H., Xing, Y., et al. (2004). A case control and family based association study of the neuregulin1 gene and schizophrenia. *J Med Genet 41*, 31-34.

Zhou, W., and Carpenter, G. (2000). Heregulin-dependent trafficking and cleavage of ErbB-4. *J Biol Chem 275*, 34737-34743.

Zhu, Y., Li, H., Zhou, L., Wu, J. Y., and Rao, Y. (1999). Cellular and molecular guidance of GABAergic neuronal migration from an extracortical origin to the neocortex. *Neuron 23*, 473-485.

Zrihan-Licht, S., Avraham, S., Jiang, S., Fu, Y., and Avraham, H. K. (2004). Coupling of RAFTK/Pyk2 kinase with c-Abl and their role in the migration of breast cancer cells. *Int J Oncol 24*, 153-159.

NOTES ADDED IN PROOF

Data referred to as Cleary et al., unpublished data, are now in press [1].

A recent paper by Sawamoto and colleagues [2] demonstrates that migration of neuroblasts around the lateral ventricles parallels the flow of cerebrospinal fluid (CSF). The beating of ependymal cilia was required for normal CSF flow, neuronal migration, and the formation of a gradient of the chemorepulsive factor slit2 (discussed above). The authors suggest that creating a gradient of slit protein in the overlying SVZ, the flow of CSF in the lateral ventricles may provide important directional information to cells before they reach the RMS.

Two recent papers by Nelson, Weiss and colleagues [3, 4] provide further insight into the role of the cadherin and catenin families of proteins in regulating actin filament assembly. In the first of these papers [4] they show that contrary to what had been believed previously, α-catenin is unable to bind simultaneously to actin and to a membrane complex containing β-catenin and E-cadherin. In the second paper [3] they show that α-catenin suppresses Arp 2/3 mediated actin polymerization (see figure 3), probably by competing with Arp 2/3 for binding to actin. Interestingly, binding of α-catenin to actin was inhibited by binding of α-catenin to a membrane complex containing β-catenin and E-cadherin.

A review of the role of extracellular matrix proteins in neuronal migration during cerebellar development has been published [5] and offers some interesting parallels and contrasts with the findings described here for the adult SVZ and RMS.

An excellent review of reelin-related signalling pathways as well as a comprehensive description of the reeler phenotype has recently been published [6]. Two recent studies also provide additional insight into the mechanism of reelin signalling in migrating neurons. Simo and colleagues [7] found that Reelin triggers ERK signaling in an SFK/mDab1- and PI3K-dependent manner and that ERK activation is required for Reelin-dependent detachment of neurons migrating from the SVZ, while Oshima and colleagues [8] showed that Cdk5 modulates Reelin signaling through the Ser/Thr phosphorylation of Dab1 in an isoform-specific manner.

An article by Wilson and colleagues [9] shows that in addition to their roles in axon guidance and neuronal migration (reviewed above) netrins also stimulate proliferation, migration and tube formation in endothelial cells.

[1] Cleary, M.A., et al., *Expression of Ezrin in Glial Tubes in the Adult Subventricular Zone and Rostral Migratory Stream.* Neuroscience, 2006. **in press**.

[2] Sawamoto, K., et al., *New neurons follow the flow of cerebrospinal fluid in the adult brain.* Science, 2006. **311**(5761): p. 629-32.

[3] Drees, F., et al., *Alpha-catenin is a molecular switch that binds E-cadherin-beta-catenin and regulates actin-filament assembly.* Cell, 2005. **123**(5): p. 903-15.

[4] Yamada, S., et al., *Deconstructing the cadherin-catenin-actin complex.* Cell, 2005. **123**(5): p. 889-901.

[5] Porcionatto, M.A., *The extracellular matrix provides directional cues for neuronal migration during cerebellar development.* Braz J Med Biol Res, 2006. **39**(3): p. 313-20. Epub 2006 Feb 22.

[6] D'Arcangelo, G., *The reeler mouse: anatomy of a mutant.* Int Rev Neurobiol, 2005. **71**: p. 383-417.

[7] Simo, S., et al., *Reelin Induces the Detachment of Postnatal Subventricular Zone Cells and the Expression of the Egr-1 through Erk1/2 Activation.* Cereb Cortex, 2006.

[8] Ohshima, T., et al., *Modulation of Reelin signaling by Cyclin-dependent kinase 5.* Brain Res., 2006. **[Epub ahead of print]**.

[9] Wilson, B.D., et al., *Netrins Promote Developmental and Therapeutic Angiogenesis.* Science, 2006. **313**(5787): p. 640-644.

In: Frontiers in Stem Cell Research
Editor: Julia M. Spanning, pp. 39-54

ISBN 978-1-60021-294-8

Chapter II

ADULT STEM CELL HOMING MECHANISM AND DELIVERY STRATEGY: ACUTE VERSUS CHRONIC MYOCARDIAL INFARCTION

C. Wan, R. Atoui, J. Luo, D. MacDonald,
B. Burstein, RC-J. Chiu and D. Shum-Tim*
Division of Cardiac Surgery, McGill University Health Center, Montreal, Quebec, Canada

ABSTRACT

Adult bone marrow stem cell therapy has been the subject of tremendous research in recent years. Adult marrow stem cells (MSCs) participate in the regeneration of the myocardium after acute ischemic injuries. This process, in part,, in part, involves migration of MSCs from remote regions to sites of injuries, as well as milieu-dependent differentiation initiated through a complex interaction of signals. This process of differentiation ultimately lead to an improvement of cardiac function. Therapeutically, various approaches to the delivery of these MSCs have been proposed. However, in order to develop the most effective delivery regimen, insights into the homing-mechanisms are crucial. These mechanisms involve a complex interaction between cellular signaling and paracrine, or endocrine factors.

In an early study, we found that acute ischemic insult provided a strong signaling mechanism that directs homing of MSCs residing in bone marrow after systemic transfusion of MSCs. This signaling mechanism was reproduced in a subsequent study to investigate the presence or absence of homing signals in a chronic myocardial injury model. Interestingly, in subjects with chronic ischemic injury, the homing of MSCs was absent. In a subsequent study, we investigated whether these homing signals in the

* Correspondence concerning this article should be addressed to Dominique Shum-Tim, M.D. Division of Cardiothoracic Surgery, The Montreal General Hospital, MUHC 1650 Cedar Avenue, Suite C9-169 Montreal, Quebec, H3G 1A4, Canada. Tel: (514) 934-1934 ext. 44326; Fax: (514) 934-8289; Email: dshumtim@yahoo.ca.

recruitment of MSCs for myocardial regeneration were present in immature hearts, and whether normal physiologic growth provided similar homing signals in the recruitment of MSCs. We found that homing signals were present even in the immature myocardium after acute injury to effect the identical regeneration process in their adult counterpart. However, normal physiologic growth provided no recruiting signals to MSCs.

Insights into the cellular and molecular basis of homing mechanisms have been shed in recent years. Implicated in the process of MSC recruitment to the injured myocardium, myocardial regeneration and angiogenesis are G-CSF, M-CSF, VEGF, SCF-1, among many others.

Although the exact interplay between various homing signals remains to be defined, the results of our studies have important implications on stem cell therapeutics in myocardial regeneration. Further studies to elucidate the homing mechanisms in acute myocardial injury will be required.

Keywords: Adult stem cells, homing signals, acute myocardial injury, animal models, myocardial regeneration, cellular cellular cardiomyoplasty

A. Background Information

Adult bone marrow stem cell therapy has been the subject of tremendous research in recent years, both experimental and clinical. After an acute ischemic injury, adult bone marrow stem cells (MSC) have been shown to participate in the regeneration of the myocardium, through a process that involves 1) migration of these stem cells from remote regions to the site of injury; and 2) subsequent differentiation into cardiomyocytes, endothelial cells and smooth muscle cells within a specific and unique microenvironment, a process that our group has coined the term “milieu-dependent differentiation” [1]. Participation of these stem cells in scar remodeling and maturation, myocardial regeneration, angiogenesis and vasculogenesis is very promising and reinforces the notion of stem cell transplantation as a potential therapy in various forms of acquired heart disease, including acute myocardial infarction [2], chronic ischemic cardiomyopathy [3] and valvular heart disease [4].

As a result of the ability of these adult stem cells to participate in myocardial repair and regeneration and their potential to ameliorate heart failure, human clinical trials for the treatment of acute myocardial ischemia have been launched with promising results [2-7]. Potential routes of delivery of stem cells in therapy include: 1) intravenous transfusion; 2) catheter-based trans-coronary transfusion; 3) catheter-based trans-coronary venous transfusion [8]; 4) catheter-based trans-endocardial injection; 5) direct trans-epicardial injection via an open, minimally invasive, or imaging-guided procedure. The potential advantages and disadvantages of each delivery method are outlined in table 1. The majority of human clinical trials have employed autologous stem cells were delivered via intracoronary or trans-endocardial meythods, although the optimal route of stem cell delivery has not yet been determined [9-12].

Table I. Various routes of delivery of stem cells

Route of Delivery	Advantages	Disadvantages	Transfusion volume
Intravenous	• Easey of use • Truly minimally invasive • Practical	• First-pass effect of the pulmonary circulation • Unknown safety • Low cell implantation rate	Largest
Intra-Coronary transfusion	• Relative easey of use • Delivery to myocardial tissues in higher concentrations than intravenous delivery	• Delivery to hypoperfused and non-perfused region difficult • Potential myocardial infarction from transfusioncoronary occlusion	Medium
Coronary venous transfusion	• Avoid manipulation and embolization to large coronary arteries • Potential avoidance of micro-infarcts	• Unknown efficacy in delivery • Potential avoidance of micro-infarcts	Medium
Trans-endocardial injection	• Targeted delivery	• Unknown rRisk of cardiac chamber perforation • Dye load	Small
Trans-epicardial injection	• Targeted delivery • Most accurate delivery to targeted regions through open or minimally invasive techniques • High concentration of cells to targeted region	• Invasiveness of procedure	Smallest

In order to gain understanding of the optimal route of delivery of stem cells for therapeutic purposes, and the consequent implication on the amount of cells required, knowledge into the characteristics and mechanisms of their migration to sites of injury is crucial. In a previous study, Chedrawy et al. [13] from our group has demonstrated that direct injection of stem cell into myocardium is an effective method of delivering mesenchymal stem cells into the myocardium. These stem cells subsequently underwent milieu-dependent differentiation into differentiated cardiomyocytes. In another study, bone marrow cells pretreated with 5-azacytidine were transplanted through trans-coronary means, to infarcted rat hearts [14]. These cells were not only found to differentiate into cardiomyocytes but to also contribute in the significant improvement in the overall cardiac function after implantation.

In the last several years, several studies have shown the importance of paracrine factors in providing the specific signals for the migration of the marrow stem cells to the targeted, infarcted regions of the heart [15]. These findings are consistent with previous studies demonstrating the participation of stem cells in the repair process after myocardial injury, both through direct delivery [16] or cytokine-induced recruitment [17].

Based on these findings, we hypothesized that specific cell signals, released at the time of an acute myocardial injury permit the recruitment and the trafficking of MSCs from the bone marrow to the site of injury in order to participate in myocardial regeneration and the potential improvement in the cardiac function.

In this chapter, we will summarize our findings and emphasize the importance of cellular signaling and paracrine factors in the recruitment process of the marrow stem cells in two different scenarios: acute and chronic myocardial infarctions. We will also briefly describe our results regarding the role of MSC in normal post-natal myocardial growth.

B. Acute Myocardial Injury Provides Signals for Marrow Stromal Cell Mobilization and Homing

Repair process takes place aAfter a acute myocardial infarction. This process is, a repair process takes place characterized by an acute inflammatory process involving scavenger cells that remove tissue debris. This is Scar formation followsed by scar formation, leading toand eventual angiogenesis, and myogenesis [18], leading toand eventual remodeling. We hypothesized that the post-infarct inflammatory response that ultimately leads to ventricular remodeling involves marrow stromal cells. Furthermore, specific homing signals are present after an acute ischemic event to help mobilizing these MSC from remote sites [19].

Methods

Animal Models

Male isogenic Lewis rats from Charles River laboratories were used. Animal care and handling were in accordance to the Canadian Council on Animal Care.

Adult Marrow-Derived Stromal Cells Collection and Culture

Marrow stromal cells were isolated from the bones of donor rats according to the Caplan's method [20]. The femoral and tibial bones were collected and both ends of the long bones were cut away from the diaphyses. The bone marrow plugs were hydrostatically expelled from the bones with complete medium, consisting of Dulbecco's Modified Eagle's Medium (DMEM) containing selected lots of 10% calf serum and antibiotics (100 U/ml penicillin G, 100 mg/mg streptomcycin, 0.25 mg amphotericin B (Gibco laboratories) in a humidified atmosphere of 5% CO_2. The marrow plugs were disaggregated and the dispersed cells were centrifuged and resuspended twice in complete medium. These cells in 10 ml of complete medium were then introduced into tissue culture dishes. Medium was replaced every 3 days and the non-adherent cells discarded. Each primary culture was twice divided into three new plates and cultured until the cell density of the colonies grew to approximately 90% confluence.

Adult Marrow Stromal Cells Labeling

LacZ-GP + AM12 amphotropic retrovirus producer cells were used to label the adult stem cells. These cells produce a replication defective retrovirus containing the LacZ gene that encodes for the bacterial β-galactosidase enzyme [21]. These cells were cultured in DMEM with 10% FBS and antibiotics (50 U/ml Penicillin G and 50 mg/ml Streptomycin (Wisent Inc.). The cells were allowed to proliferate until at least a 70% confluence was achieved before using them for transfection in order to achieve a high retrovirus titer. Twenty-four hours before transfection, the marrow stromal cells were trypsinized with 0.05% Trypsin + 0.53 mM EDTA (Gibco Laboratories) and replated. The next day, these cells were cultured with the filtered supernatant from retrovirus-producing Lac-Z-GP AM12 cells and Polybrene (4microgram in 1 ml of DMEM). Seventy-two hours after the last transfection, marrow stromal cells were trypsinized and part of the cells were plated in a 35-mm dish for histochemical staining for β-galactosidase activity in order to determine the percentage of cells expressing β-gal activity. The medium was aspirated from the plates and the cells rinsed with phosphate buffered saline (PBS). The cells were fixed at 4°C in fix solution (2% formaldehyde and 0.2% glutaraldehyde in PBS) for 15 min and re-rinsed with PBS. Staining for β-gal activity was performed with a solution containing 1 mg/ml 5-bromo-4-chloro-3-indoyl-b-D-galactoside (X-gal), 2% dimethylsulfoxide, 20 mM K3Fe(CN)6, 20 mM K4Fe(CN)63H2O and 2 mM magnesium chloride.Staining for β-gal activity was performed with solution containing 2% 5-bromo-4-chloro-3-indoyl-b-D-galactoside (X-gal) in dimethylFormamide, 20mM ferricyanide, 20mM ferrocyanide and 2mM magnesium chloride. The cells were then incubated at 37°C, pH 7.8 and protected from light for 16 hrs. The presence of blue-labeled cells was then confirmed under phase microscopy.

Stromal Cell Suspension for Injection

Cells isolated from bone marrow were cultured in complete medium in tissue culture dishes. The medium was aspirated and the cells in each dish were washed with 6 ml of Hank's Basic Salt Solution (HBSS). The HBSS was aspirated and 2 ml of trypsin-EDTA was added to detach the cells from the bottom of the dish. The detached cell suspension was then placed in a flask with 2 ml of complete medium and placed in a hematocytometer for counting; 15 x 10^{76} cells was then collected and centrifuged at 2500 rpm for 5 min, and the cell pellet resuspended in 0.5 ml of complete medium.

Marrow Stromal Cell Transplant

Marrow stromal cell injection was carried out by intravenous route through the peripheral vein. The presence of labeled marrow stromal cells was confirmed in recipient rats at various time points ranging from 2 hours to 6 weeks. Bone marrow stromal cells were harvested and cultured as in donor rats and stained for β-galactosidase activity as described above.

Acute Myocardial Injury vs. Sham Operated Subjects

One week after the original injection of marrow stromal cells, acute myocardial injury (AMI) was produced with ligation of the left coronary artery. Animal subjects were anesthetized with isoflurane, and a left thoracotomy was performed. The left coronary artery was identified, exposed, and a 7-0 polypropylene suture was used to ligate the coronary artery. A similar operation was performed on sham-operated animal except that ligation of the left coronary artery was omitted.

Histological Evaluation

Marrow stromal cell transplant recipients were sacrificed at various time points: 1 week (n=3); 2 weeks (n=3); 4 weeks (n=6); 6 weeks (n=5); 8 weeks (n=3). The hearts were harvested and rinsed with PBS and perfusion fixed in 2% paraformaldehyde in PBS. The staining for β-galactosidase activity was performed as described above, but with the addition of 0.02% Nonidet P-40 and 0.01% deoxycholate to the staining solution. The gross heart specimens were stained for 16 hr at 37°C and pH 7.8. After X-gal staining, they were embedded in paraffin and coronal sections 5 μm thick were mounted on a set of gelatin coated glass slides such that serial sections could be used for different stains. A series of sections from each heart specimen were stained with hematoxylin and eosin and another series stained with Picrosirius Red stain for connective fibrous tissue. Other serial sections from each heart were used for immunohistochemical stainings, for sarcomeric myosin heavy chain molecules with MF20 (Developmental Studies Hybridoma Bank developed under the auspices of the NICHD and maintained by the University of Iowa, Department of Biological Sciences); for connexin 43 (Zymed Laboratories Inc, San Francisco, CA); for troponin I-C (Santa-Cruz Biotechnology Inc); and for alpha smooth muscle actin (Sigma Laboratories). Briefly, after de-paraffinization, sections were placed in boiled citrate buffer (pH 6.0). After blocking in normal serum, sections were treated with the respective monoclonal antibodies overnight and with secondary antibody the following day. Diaminobenzidine (DAB), which produced brown color, was then used as a chromogen for light microscopy. Counter-staining of sections by hematoxylin and eosin was also performed. Cells derived from labeled marrow stromal cells were identified by their blue nuclei.

Results

Marrow Stromal Cell Transplantation

Intravenous implantation of cells yielded no mortality and thewas well tolerated by the rats animal subjects.tolerated the procedure well The homing of labeled cells into the bone marrow of the recipient rats after implantation was confirmed at various time points after Xx-gal staining. In all rats, both in the sham-operated and coronary artery ligationed groups, the harvested bone marrow contained 1–2% labeled cells within each culture plate.

Acute Myocardial Injury vs. Sham

Gross examination of the infarcted hearts in the coronary artery ligation group revealed a scars in the distribution of the left coronary artery in all hearts, centered on in the anterior aspect of the left ventricle. After X-gal staining, a blue discoloration was localized in the area of scarred myocardium in all the coronary ligated hearts (figure 1). The sham-operated hearts also underwent X-gal staining, which revealed the absence of blue color in any of the hearts.

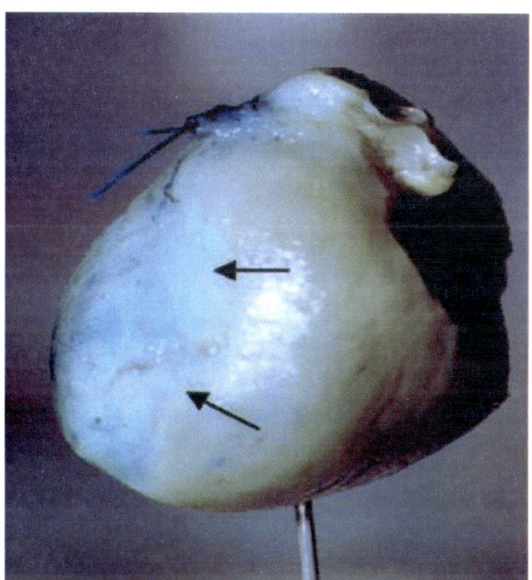

Figure 1. Staining of the myocardium with Xx-gal is limited to the peri-infarcted regions.

When serial coronal cross-sections of the hearts were examined, labeled marrow stromal cells were identified in all 14 coronary ligated hearts in the peri-infarted region, and in none of the 6 sham-operated hearts. Up until 3 weeks post-infarct, the MSCs showed an immature monocyte-like appearance with a large nucleus to cytoplasm ratio. Heart specimens obtained and studied at all earlier time points did not reveal any clear evidence of differentiation. However, 3 weeks after coronary artery ligation, immunohistochemical staining showed evidence of cardiomyogenic differentiation in some labeled cells with the presence of the cardiomyocyte specific protein troponin I-C [22]. The staining for troponin I-C was more pronounced in the cytoplasm of these cells by 4–6 weeks, with further morphologic changes indicating the gradual differentiation of the MSC into fully mature cardiomyocytes. Six weeks after implantation, the morphology of the implanted cells was clearly different from the one seen at 1 week. They appeared more elongated in shape, aligned with adjacent labeled cells and were better more able to integrated within the host cardiomyocytic network (figure 2). Furthermore, as early as 4 weeks after implantation, these cells stained positive for connexin-43, a main constituent protein in the cardiac gap junctions [23] was evident. Finally, further specific staining confirmed our milieu-dependent differentiation during which MSC differentiated into not only cardiomyocytes, but also into endothelial cells, lining the walls of small blood vessels as well as into smooth muscle cells [24].

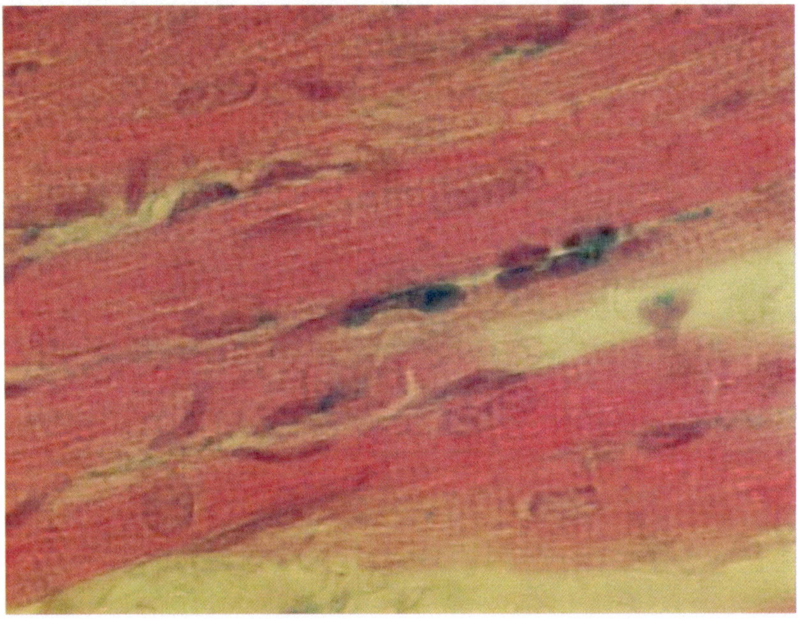

Figure 2. Labeled marrow stromal cells aligned with native myocardial cells in the direction of their fibers.

Interpretation of Results

In this study, we have shown that MSCs transfused intravenously in a rat model homed into the bone marrow, as can be demonstrated by the presence of labeled MSCs from the long bones of the recipient animals. The resultant subsequent acute injury to the myocardium must provided important signals to the marrow stromal cells to migrate from remote regions within the bone marrow to reach the injured site. These marrow stromal cells then underwent a milieu-dependent differentiation into cardiomyocytes within the peri-infarct regions, as suggested by their expression of cardiomyocyte-specific markers. In addition, these differentiated cells showed signs that they are inter-connected and aligned with the remaining cardiomyocytes. Although functional performance assessment was not carried outperformed in this series of experimentstudy, we believe this sequence of events will translate into a functional improvement, as previous studies have demonstrated that this process contributed to meaningful functional contribution, improvingimprovement on not only the overall cardiac function, but attenuation of ing as well the negative remodeling process that accompanies a myocardial infarction [25].

C. Acute *vs* Chronic Ischemia

While acute myocardial infarction (AMI) provided strong homing signals to the implanted marrow stromal cells, it is unknown whether the same holds true for chronic ischemia (CMI). In the subsequent studies, we hypothesized that in chronic ischemia, these paracrine homing signals allowing MSC mobilization are absent [26].

Methods

Animal Models

Male isogenic Lewis rats from Charles River laboratories were used. Animal care and handling were in accordance to the Canadian Council on Animal Care.

Adult Marrow-Derived Stromal Cells Collection and Culture

Marrow stromal cells were isolated from the femoral and tibial bones of donor rats according to the Caplan's [20] method as described above.

Adult Marrow Stromal Cells Labeling

Lac-Z gene that encodes for the bacterial ß-galactosidase enzyme was transfected in MSCs using same methodology as in the previous experimentstudy. The presence of blue-labeled cells was then confirmed under phase microscopy.

Coronary Artery Ligation

A total of 20 male Lewis rats were assigned to acute myocardial infaction (AMI) or chronic myocardial infarction (CMI) groups. Left coronary artery ligation was performed through a left thoracotomy under general anesthesia in each experimental subject.

Marrow Stromal Cell Transplantation

In the AMI group, isogenic MSCs (10^7 cells) labeled with Lac-Z reporter gene were injected intravenously through a peripheral vein within 24 hours after coronary ligation. In the CMI group, similar amount of labeled MSCs were given 6 weeks after coronary ligation.

Histologic Examination

Hearts were explanted and histologically examined with X-gal staining to identify the presence of labeled cells at 1 week and 3 weeks after MSCs injection. If positively labeled cells were observed, the extent of differentiation was analyzed with with immunohistochemical staining targeting for cardiac-specific markers such as troponin I-C and connexin-43.

Results

Coronary Artery Ligation

All the rats subjected for coronary ligation (n=20) survived the operation and subsequent MSCs injection.

One Week Post-Marrow Stromal Cell Transplantation

At 1 week post-marrow stromal cell transplantation, half of the hearts (n=5/group in the AMI and CMI groups) were harvested to examine the presence of Lac-Z MSCs. Explanted hearts were stained with X-gal solution and gross examination revealed blue discoloration localized in the area of specimens taken from the AMI group. No blue discoloration was detected in any heart harvested from the CMI group. This finding was confirmed after H&E staining, where no lac-Z positive cells were identified within the myocardium of the CMI group.

Three Week Post-Marrow Stromal Cell Transplantation

The paucity of MSCs in CMI animals persisted at 3-week time point. In contrast, all the hearts in the AMI group were found to stain positively with X-gal solution in the peri-infract region of the myocardium indicating the presence of Lac-Z labeled MSCs, as demonstrated in our previous experiment. Furthermore, some of the labeled MSCs were found to align with native myocardial fibers and to stain positive for troponin I-C.

Interpretation of Results

In this study, we investigated whether a chronic myocardial injury can provide specific homing signals to mobilize MSCs into the injured site as it was seen in the acute ischemic model. Marrow stromal cells were transfused intravenously either shortly after coronary ligation (AMI model) or 6 weeks after the infarction (CMI model). A recruitment process much like what we have previously observed in the prior study was evident only in the AMI model where the intravenously transfused marrow stromal cells were found to home to the peri-infarct regions of the myocardium and to gradually participate in myocardial regeneration through differentiation into mature cardiomyocytes. However, this finding was not mirrored in the chronic myocardial infarction group where there was no apparent homing of the injected MSCs at any time throughout the study period from 1 to 3 weeks. This lack of signaling mechanism in the CMI may haveprovides important clinical implication when designing strategies for cell therapy in patients with myocardial infarction.

D. The Role of Marrow Stromal Cells in Acute Myocardial Injury of the Immature Heart Normal Growth

In order to better assess the nature and importance of homing signals involved in MSCs recruitment, we hypothesized that growth, as in acute injury, may provide specific signals for adult MSCs, to be recruited and to participate in normal physiologic growth of myocardial tissues. We also hypothesized that acute ischemic injury to the immature myocardium provides the same homing mechanisms as seen in mature hearts.

In a recent study, the same donor MSCs were injected systemically into groups of neonatal (NB=2days-old), immature (B=30days-old) and adult (A= >3months-old) isogeneic Lewis rats (n=10/group). Additionally, left coronary artery (LCA) ligation was performed in a subgroup of the immature (BL) and adult (AL) rats one week after MSCs injection. The hearts were harvested serially from 2 days to 6 weeks and stained with X-Gal for labeled MSCs. Cardiomyocyte phenotypic expression was evaluated as previous studies using immunohistological staining for troponin I-C and connexin-43.

Results

Labeled MSCs were found to home into bone marrow of all the recipients at all ages, but not in the normal hearts of any neonatal, growing rats or adult rats. However, as seen in the previous experiments, the labeled stromal cells were recruited from the bone marrow into the infarcted site of myocardium only after ligating the left coronary artery in groups AL and BL. In addition, these recruited stromal cells were capable of differentiating into cardiomyocyte phenotype after myocardial injury. In contrast to that reported in the developing fetus [27], MSCs did not appear to contribute to the growth of non-injured hearts from neonates through adult ages in groups A, B and NB.

Interpretations of Results

Marrow stromal cells did not appear to contribute to the growth of non-injured hearts from neonates through adult ages in groups A, B and NB. However, they were capable of regenerating damaged myocardium both in the adult and in the immature hearts, presumably secondary to specific signals that appeared after creating the acute ischemic event. This is in contrast to that reported in the developing fetus in which MSCs were found to survive and underwent in situ differentiation in various organs and formed a stable chimera during fetal development [27]. This unique role of stem cells in fetal development seems to differ from that after birth when organogenesis in most organs ceases. In addition, acute injury in immature myocardium effects the same sequence of events that ultimately lead to engraftment, differentiation and integration of MSCs to the myocardium.

E. Discussion

In our first study, we have shown that acute myocardial injury mobilized transplanted isogenic adult MSCs residing in the bone marrow and trafficked through the circulation into the acutely injured myocardium. They engrafted in the peri-infarct regions of the myocardium and subsequently underwent differentiation into cardiomyocytes, expressing cardiac-specific markers, such as troponin I-C and connexin-43. This confirmed the presence of specific homing signals targeting these MSCs after organ injury. This phenomenon sheds important insights into the potential clinical applications of stem cell therapeutics. If a less invasive method of cell delivery through intravenous administration of cell therapy is considered, it should probably be given during early period after acute myocardial injury as homing signals for the transplanted stromal cells are still present.

In a subsequent study, we have demonstrated that only acute injury, as opposed to a chronic insult. provided the homing signals for the intravenously transfused MSCs. Transfused MSCs would homed into the region of the acute injury, underwentgo differentiation and integrated themselves into the peri-infarct regions only if given early after AMI. These studies are of paramount importance as they suggest that a systemic administration of MSC may not be efficient effective in a chronic ischemic model. Rather, it would probably requireConsideration should be given to a direct local delivery of therapeutic stem cells, either via intracoronary injection, trans-endocardial or trans-epicardial injection of stem cells to the targeted areas.

Our third study confirmed our previous finding related to MSC mobilization. It seems that specific signals, involved in the recruitment of MSC, appear only after the creation of an acute ischemic event regardless of the age group of the recipients. It appears then that these cells seem to have a natural role in repair and remodeling of an acute myocardial injury. They do not appear to play a significant role in chronic insults, or in natural post-natal growth. These findings are confirmed in several other studies in which MSC differentiation appeared to occur almost exclusively in organs that have been damaged acutely. For instance, differentiation to endothelial cells, hepatocytes, and myoblasts is seen in cases models of ischemia, cirrhosis and muscular dystrophy [28,29,30]. In these reports, it is suggested that upon injury, stem cells can proliferate in vivo and are then recruited to the injured environment by specific signals.

The molecular basis for homing of adult MSCs into injured myocardial tissue has received a lot of attention in recent years, and is currently under active investigation [31,32,33]. Granulocyte colony stimulating factor (G-CSF) and macrophage colony stimulating factor (M-CSF) have been shown to mobilize bone marrow cells in order to participate in the repair process after myocardial injury [34,35,36]. Tang et al [15] highlighted the importance of paracrine action of the engrafted MSC in the ischemic myocardium and the resulting stimulation of neovascularization. They also showed that the release of bFGF, VEGF and SDF-1 did not only leads to efficient vascular regeneration but also attenuateds the apoptotic pathway by downregulating the prosurvival protein Bax. Finally, it was also shown that local injection of MSC-derived conditioned media alone containing contained several arteriogenic cytokines that could enhance collateral perfusion in a murine model of hindlimb ischemia, hence highlighting the important role of paracrine signaling [37].

Another recent study demonstrated that stromal cell-derived factor-1α (SDF-1) was an important player in the recruitment of MSCs to the heart [38]. Interestingly, in this study, SDF-1 provided strong homing signals to the MSCs only in the presence of an acute ischemic injury, as observed in our study. It is of interest to note that although SDF-1 was found to be required in stem cell mobilization to the injured site, it is not singularly sufficient, hence reflecting the need for additional factors. In fact, patients with acute coronary syndromes have elevated levels of many factors other than SDF-1 including MMP-2, MMP-9, ICAM, and VCAM [38].

Although the exact mechanism of MSCs mobilization remains undetermined, it is believed to be a multi-step process whereby a key process involves the disruption of the adhesion interactions between the stem cells and the bone marrow when stimulated by specific signals such as an ischemic injury or stress. In fact, Papayannopoulou et al. demonstrated the critical role of VLA-4 in the mobilization process [39] and several previous studies have shown the role of proteolytic enzymes such as elastases, peptidases, cathepsins G, MMP-2 and MMP-9 in inactivating SDF-1 by cleaving part of its N-terminus. Furthermore, a recent study showed that the increase in the level of cathepsin G and elastase correlate with stem cell mobilization [40]. In accordance with these studies, Petit and her group demonstrated the proteolytic degradation of SDF-1 by elastase induced by G-CSF, accompanied by a gradual increase in CXCR4 expression on bone marrow cells [41]. Askari et al. further reinforced the role of SDF-1 in stem cell homing in a study whereby cardiac fibroblasts expressing SDF1 were transplanted into the infarcted regions of rat hearts [42]. After using G-CSF to mobilize stem cells, a significant homing of c-kit cells to the injured area as well as an improved cardiac function was found in treated animals.

The recognition of the broad growth and differentiation potential of MSCs has opened the door to several many clinical applications. The importance of cell signaling, and paracrine and endocrine factors involving the interplay of various cytokines and growth factors remain to be better understood in order to unveil new areas of clinical opportunities in stem cell therapy. It is therefore clear that in spite of the great promise of stem cell cardiomyoplasty, many challenges remain to be solved. Investigations at both experimental and clinical levels are being pursued, and it is hoped that this fascinating therapeutic approach could prove to be a potent therapeutic tool for myocardial regeneration.

REFERENCES

[1] Chiu RC. Bone-marrow stem cells as a source for cell therapy. *Heart Failure Reviews. 8(3)*:247-51, 2003.

[2] Assmus B. Schachinger V. Teupe C et al. Transplantation of Progenitor Cells and Regeneration Enhancement in Acute Myocardial Infarction (TOPCARE-AMI). *Circulation. 106(24)*:3009-17, 2002.

[3] Perin EC. Dohmann HF. Borojevic R et al. Transendocardial, autologous bone marrow cell transplantation for severe, chronic ischemic heart failure. *Circulation. 107(18)*:2294-302, 2003.

[4] Perry TE. Kaushal S. Sutherland FW et al. Thoracic Surgery Directors Association Award. Bone marrow as a cell source for tissue engineering heart valves. *Annals of Thoracic Surgery. 75(3)*:761-7

[5] Strauer BE. Brehm M. Zeus T et al. Repair of infarcted myocardium by autologous intracoronary mononuclear bone marrow cell transplantation in humans. *Circulation. 106(15)*:1913-8, 2002.

[6] Wollert KC. Meyer GP. Lotz J et al. Intracoronary autologous bone-marrow cell transfer after myocardial infarction: the BOOST randomised controlled clinical trial.[see comment]. *Lancet. 364(9429)*:141-8, 2004.

[7] Galinanes M, Loubani M, Davies J, et al. Autotransplantation of unmanipulated bone marrow into scarred myocardium is safe and enhances cardiac function in humans. *Cell Transplant 2004*; 13: 7–13.

[8] Thompson CA. Nasseri BA. Makower J et al. Percutaneous transvenous cellular cardiomyoplasty. A novel nonsurgical approach for myocardial cell transplantation. *Journal of the American College of Cardiology. 41(11)*:1964-71, 2003.

[9] Aviles FF. San Roman JA. Garcia Frade et al. Intracoronary stem cell transplantation in acute myocardial infarction. *Revista Espanola de Cardiologia. 57(3)*:201-8, 2004.

[10] Kang HJ. Kim HS. Zhang SY et al. Effects of intracoronary infusion of peripheral blood stem-cells mobilised with granulocyte-colony stimulating factor on left ventricular systolic function and restenosis after coronary stenting in myocardial infarction: the MAGIC cell randomised clinical trial. *Lancet. 363(9411)*:751-6, 2004.

[11] Ozbaran M. Omay SB. Nalbantgil S et al. Autologous peripheral stem cell transplantation in patients with congestive heart failure due to ischemic heart disease. *European Journal of Cardio-Thoracic Surgery. 25(3)*:342-50.

[12] Li TS, Hamano K, Hirata K et al. The safety and feasibility of the local implantation of autologous bone marrow cells for ischemic heart disease. *J Card Surg* 2003; 18 (suppl 2):69–75.

[13] Chedrawy EG. Wang JS. Nguyen DM et al. Incorporation and integration of implanted myogenic and stem cells into native myocardial fibers: anatomic basis for functional improvements. *Journal of Thoracic & Cardiovascular Surgery. 124(3)*:584-90, 2002.

[14] Saito T. Kuang JQ. Lin CC. et al. Transcoronary implantation of bone marrow stromal cells ameliorates cardiac function after myocardial infarction. [Journal Article] *Journal of Thoracic & Cardiovascular Surgery. 126(1)*:114-23, 2003.

[15] Tang Y, Zhao Q, Qin X et al. Paracrine action enhances the effects of autologous mesenchymal stem cell transplantation on vascular regeneration in rat model of myocardial infarction. *Annals of Thoracic Surgery*, 2005; 80: 229-237

[16] Orlic D. Kajstura J. Chimenti S et al. Bone marrow cells regenerate infarcted myocardium. *Nature. 410(6829)*:701-5, 2001.

[17] Orlic D. Kajstura J. Chimenti S et al. Mobilized bone marrow cells repair the infarcted heart, improving function and survival. *Proceedings of the National Academy of Sciences of the United States of America. 98(18)*:10344-9, 2001.

[18] Hakuno D. Fukuda K. Makino S et al. Bone marrow-derived regenerated cardiomyocytes (CMG Cells) express functional adrenergic and muscarinic receptors. *Circulation. 105(3)*:380-6, 2002.

[19] Bittira B. Shum-Tim D. Al-Khaldi A et al. Mobilization and homing of bone marrow stromal cells in myocardial infarction. *European Journal of Cardio-Thoracic Surgery. 24(3)*:393-8, 2003.

[20] Caplan AI. Mesenchymal stem cells. Journal of Orthopaedic Research. 9(5):641-50, 1991.

[21] Jaalouk DE. Eliopoulos N. Couture J et al. Glucocorticoid-inducible retrovector for regulated transgene expression in genetically engineered bone marrow stromal cells. *Human Gene Therapy. 11(13)*:1837-49, 2000.

[22] Adams JE 3rd. Bodor GS. Davila-Roman VG. Delmez JA. Apple FS. Ladenson JH. Jaffe AS. Cardiac troponin I. A marker with high specificity for cardiac injury. *Circulation. 88(1)*:101-6, 1993.

[23] Beyer EC. Paul DL. Goodenough DA. Connexin43: a protein from rat heart homologous to a gap junction protein from liver. *Journal of Cell Biology. 105(6 Pt 1)*:2621-9, 1987.

[24] Wang JS, Shum-Tim D, Galipeau et al. Marrow stromal cells for cellular cardiomyoplasty : feasibility and clinical advantages. *Journal of Thoracic and cardiovascular surgery*, 2000; 120: 994-1006

[25] Davani S, Marandin A, Mersin N et al. Mesenchymal progenitor cells differentiate into an endothelial phenotype, enhance vascular density, and improve heart function in a rat cellular cardiomyoplasty model. *Circulation* 2003; 108(Suppl 1):II253-258

[26] MacDonald DJ, Burstein B, Luo J, Duong M, Shum-Tim D, Chiu RCJ.: Therapeutic Delivery Strategies of Adult Stem Cells for Acute vs Chronic Myocardial Infarctions: The Role of Homing Mechanisms. *Abstract presented at the International Society of Stem Cell Research*, Boston, June 2004.

[27] Liechty KW. MacKenzie TC. Shaaban AF et al. Human mesenchymal stem cells engraft and demonstrate site-specific differentiation after in utero transplantation in sheep. *Nature Medicine. 6(11)*:1282-6, 2000.

[28] Gussoni E., Soneoka Y., Strickland CD. et al. Dystrophin expression in the MDX mouse restored by stem cell transplantation. *Nature* 1999; 401:390-4

[29] Lagasse E, Connors H, Al-Dhalimy M et al. Purified hematopoietic stem cells can differentiate into hepatocytes in vivo. *Nat Med* 2000; 6:1229-1234

[30] Theise ND, Nimmakayalu M, Gardner R et al. Liver from bone marrow in humans. *Hepatology* 2000; 32:11-16

[31] Sorger JM, Despres D et. al. MRI tracking of intravenous magnetically labeled mesenchymal stem cells following myocardial infarction in rats. *Circulation* (2002) 106 (Suppl.):II-16.

[32] Fujita J, Suzuki Y, Ando K et. al. Transplanted bone marrow stem cells move to the infracted myocardium and differentiate into smooth muscle cells and cardiomyocytes. *Circulation* (2002) 106 (Suppl.):II-132.

[33] Terrovitis J. Charitos C. Dolou P et al. No effect of stem cell mobilization with GM-CSF on infarct size and left ventricular function in experimental acute myocardial infarction. *Basic Research in Cardiology*. 99(4):241-6, 2004.

[34] Miki T, Sakamoto J, Nakano A et. al. Mobilization of bone marrow cells by G-CSF/M-CSF improves ventricular function in the heart undergoinging postinfarct remodeling. *Circulation* (2002) 106 (suppl.):II-15.

[35] Minatoguchi S, Takamura G, Chen XH et. al. Myocardial infarction itself induces cardiomyocyte regeneration from bone marrow cells and post-ischemic G-CSF treatment improves cardiac dysfunction via acceleration of the process. *Circulation* (2002) 106 (suppl.):II-132.

[36] Fukuhara S, Tomita S, Ohtsu Y et. al. G-CSF promoted bone marrow cells to migrate into infracted heart and differentiate into cardiomyocytes. *Circulation* (2002) 106 (suppl.):II-376.

[37] Kinnaird T, Stabile E, Burnett MS et al. Marrow-derived stromal cells express genes encoding a broad spectrum of arteriogenic cytokines and promote in vitro and in vivo arteriogenesis through paracrine mechanisms. *Circ Res* 2004; 94:678-85

[38] Abbott JD. Huang Y. Liu D et al. Stromal cell-derived factor-1alpha plays a critical role in stem cell recruitment to the heart after myocardial infarction but is not sufficient to induce homing in the absence of injury. *Circulation*. 110(21):3300-5, 2004.

[39] Papayannopoulou T, Nakamoto B. Peripheralization of hematopoietic progenitors in primates treated with anti-VLA4 integrin. *Proc Natl Acd Sc U.S.A* 1993; 90:9374

[40] Levesque JP, Takamatsu Y, Nilsson SK, et al. Vascular cell adhesion molecule-1 (CD106) is cleaved by neutrophil proteases in the bone marrow following hematopoeitc progenitor cell mobilization by G-CSF. *Blood* 2001; 98:1289

[41] Petit I, Szyper-Kravitz M, Nagler A et al. G-CSF induces stem cell mobilization by decreasing bone marrow SDF-1 and upregulating CXCR4. *Nature Immunol* 2002; 3:687

[42] Askari AT, Unzek S, Popovic ZB et al. Effect of SDF1 on stem cell homing and tissue regeneration in ischemic cardiomyopathy. *Lancet* 2003; 362: 697

In: Frontiers in Stem Cell Research
Editor: Julia M. Spanning pp. 55-117

ISBN 1-60021-294-8

Chapter III

Embryonic Stem Cells: Establishment, Maintenance, and Differentiation

*Hossein Baharvand**
Department of Stem Cells, Royan Institute, Tehran, Iran.

This chapter is dedicated to Dr. Saeid Kazemi Ashtiani, a great stem cell biologist, a wonderful colleague, and an inspirational advocate of human stem cell research in Iran.

Abstract

Embryonic stem cells (ESCs) are undifferentiated cells, which are typically derived from the inner cell mass of a blastocyst-stage embryo. In culture, these cells have the capacity to self-renew in an undifferentiated state but also may differentiate into cell types representing the three embryonic germ lineages, thus revealing their pluripotent potential. These cells could be of value for creating in-vitro culture systems and animal models that may be used to study human genetic disease, gene function, drug discovery, developmental biology, and differentiation events. This review summarizes different techniques for the derivation and maintenance of undifferentiated ESCs, in particular with emphasis on human ESCs. In the second part, issues concerning self-renewal and pluripotency are explored. In addition, accumulated data regarding the in vitro differentiation potential of ESCs is evaluated. The last part of this review outlines basic criteria, which may result in the use of ESCs in future clinical applications.

Keywords: Embryonic stem cells, differentiation, establishment

* Correspondence concerning this article should be addressed to Dr. Hossein Baharvand (Ph.D), Department of Stem Cells, Royan Institute, P.O. Box: 19395-4644, Tehran, Iran. Fax: 0098-21-22414532; TEL: 0098-21-22172330; E-mail: Baharvand50@yahoo.com.

Introduction

Stem cells are defined as cells with both self-renewal and multilineage differentiation characteristics (Watt and Hogan, 2000). However, the capacity to self-renew is not the definitive behavior of stem cells, as these cells may also be non-dividing or slowly-dividing in nature. Some stem cells can be induced to proliferate upon tissue injury to achieve functional restitution. Between the stem cell and its terminally differentiated progeny there is usually an intermediate population of committed progenitors with limited proliferative capacity and restricted differentiation potential; these are the cells that display greatest proliferative activity under physiological and pathophysiological circumstances. In general, stem cells can be divided into two broad categories, depending on their origin: adult (somatic) stem cells and embryonic stem (ES) cells. Both types share the capability to produce identical daughter cells (self-renewal), and produce daughters that are fated to differentiate (Watt and Hogan, 2000). Whilst the former means that they can be expanded and used in large-scale production, the latter indicates that they can differentiate to produce desired cell types, either in vivo or in vitro.

In terms of potentiality, stem cells can be classified as totipotent, pluripotent and multipotent. Totipotency is the ability to form all cell types of the conceptus, including the entire fetus and placenta. Such cells have an unlimited capacity to give rise to any cell that may form in the adult organism. In mammals, the fertilized egg, zygote and the first 2, 4, 8, and 16 blastomeres from the early embryo, are examples of totipotent cells. Proof that these cells are totipotent arises following the observation that identical twins can be produced from splitting apart a 2-cell embryo and culturing these blastomeres individually. However, the expression 'totipotent stem cell' is perhaps a misnomer because the fertilized egg and the ensuing blastomeres from early cleavage events cannot divide to make more of them. Although these cells have the potential to give rise to the entire organism, they do not have the capability to self-renew and, by strict definition, totipotent cells of the early embryo should not be called stem cells. Pluripotency is the ability to differentiate into several cell types which represent cells derived during embryonic gastrulation and comprises of three distinct germ layers (ectoderm, mesoderm and endoderm), but not the whole organism (e.g., embryonic stem cells). Multipotent cells can form a small number of cells/tissues that are usually restricted to a particular germ layer origin (e.g. bone marrow stromal or mesenchymal stem cells) (Slack, 2000). This classification of stem cell potency is not rigid. Indeed, it is evident that the distinction between pluripotent and multipotent is becoming increasingly more blurred with some cells having greater plasticity than previously realized. What determines stem cell potency is dependent to a large extent on the genetic make-up of the cell and whether it contains the appropriate genetic programme to make a particular cell type. However, this is somewhat 'offset' by the influence of the environment in which the stem cell is placed. For example, changes in cytokine gradients, cell–cell and cell–matrix contacts are important in switching 'on' and 'off' genes and gene pathways, thereby controlling the type of cell(s) that are generated.

Multipotent adult stem cells are located in somatic tissues and are known to maintain and regenerate tissues and organ systems. The cellular basis of these properties is the capacity of adult stem cells to self-renew and to recruit the various cell types of the tissue. In adults, a highly regulated process of stem cell self-renewal and differentiation may help to sustain tissues with high cell turnover. Adult stem cells have been isolated in highly characterized tissues including bone marrow (De Haan, 2002), skin (Watt, 2001), intestinal epithelium (Potten, 1998) or liver (Theise et al., 1999), as well as in the retina (Tropepe et al., 2000), central nervous system (review Okano, 2002), pancreas (Ramiya et al., 2000) and skeletal muscle (Seale et al., 2001). The proliferative capacity of these cells in vivo is not evident. Although adult stem cells have been found to be more versatile than originally believed (Krause et al., 2001), they typically can differentiate into a relatively limited number of cell types.

Embryonic stem cells (ESCs) are derived from the inner cell mass of the blastocyst and have an exhibit high proliferative capabilities and upon induction of differentiation, may form virtually any specialized tissue. However, because of their embryonic origin, ESCs differ from other stem cells in their ability to retain the potential to generate derivatives of all three germ layers. Embryonic stem cells have since been shown to contribute to all cell lineages, including the germ line, following microinjection studies in murine embryos which give rise to chimeras (Bradley et al., 1984; Nagy et al., 1990). In vitro, murine ESCs can be propagated indefinitely in an undifferentiated state, but also retain the capacity to differentiate into all mature somatic phenotypes upon specific induction cues. Control of ESC differentiation into specific cell types which can be maintained and cultured as pure populations, may provide an unlimited source of cells for transplant in the treatment of several diseases. An important development in this field stemmed from establishment of human ESCs (Thomson et al., 1998). The unique properties of ESCs suggest they will be of great use in clinical applications such as transplantation and tissue regeneration therapies, as well as for studying basic developmental biology, cancer research, and intensive drug and toxic screening, as an alternative method for animal assays.

This chapter reviews the current status of ESC research, with particular emphasis on progress towards the development of derivation, maintenance, differentiation, and therapies for human disease.

Pluripotent Stem Cells

Pluripotent stem cells have the ability to form any cell type from the embryonic germ layers (ectoderm, mesoderm and endoderm). In theory, pluripotent stem cells have the ability to form all the 200 or so cell types in the body. There are four classes of pluripotent stem cells. These are embryonic carcinoma cells (ECCs), embryonic germ cells (EGCs), embryonic stem cells (ESCs), and recently the discovery of a fourth class of pluripotent stem cell, the multipotent adult progenitor cells (MAPCs) from bone marrow, however, only mouse and chicken ESCs have been shown to form germ cells.

In 1970, early mouse embryos grafted into adult mice were found to produce teratocarcinomas (Solter et al., 1970; Stevens, 1970). Teratocarcinomas are malignant multidifferentiated tumors containing a significant population of undifferentiated cells. Previous studies investigating spontaneously occurring teratocarcinoma indicated that the undifferentiated component, embryonal carcinoma, could be propagated in culture. This also occurs in embryo-derived teratocarcinoma. The derivative ECC lines can be expanded continuously but can also differentiate either in vitro or via teratocarcinoma formation. Most importantly, clonally isolated and expanded ECCs retains the capacity for differentiation and may produce derivatives of all three embryonic germ layers (Kleinsmith and Pierce, 1964; Martin and Evans, 1975). Therefore, individual ECCs can be defined as self-renewing, pluripotent stem cells. The presence and expansion of ECCs in embryo-derived teratocarcinomas suggest that the epiblast founder cells of the embryo (Hogan et al., 1994), which are normally present only transiently, may in fact be predisposed for conversion to ECCs and thus for continuous self-renewal. The epiblast origin of the ECCs was evidenced by the fact that teratocarcinomas can only be produced from embryos prior to gastrulation and from grafts containing epiblast, and not from other regions of the early embryo (Diwan and Stevens, 1976). Consistent with this, ECCs show a range of phenotypic features in common with epiblast cells (reviewed by Martin, 1980). When this relationship was subjected to the ultimate test and reintroduced into the developing embryo (Brinster, 1974), some ECC lines were found to participate in embryogenesis and contribute to the formation of different tissues in the resultant chimeric fetuses and in liveborn mice. Thus, ECCs can remain receptive to cues in the embryo microenvironment and in an equivalent manner to resident epiblast cells. However, this proved to be the exception rather than the rule. Most ECC lines have poor differentiation potential in vitro and in vivo and contribute poorly to chimeras and/or produce embryonic tumors. Moreover, ECCs are almost always aneuploid, which is likely due to the uncontrolled selection pressures during tumor growth. Consequently, they are not capable of proceeding through meiosis to produce mature gametes.

Teratocarcinomas also develop spontaneously from primordial germ cells in some mouse strains, or following transplantation of primordial germ cells into ectopic sites. In 1992, Brigid Hogan and her colleagues reported the direct derivation of EGCs from mouse primordial germ cells (Matsui et al., 1992). Primordial germ cells (PGCs) are diploid germ cell precursors that transiently exist in the embryo before they migrate and form a close association with the somatic cells of the gonad, whereby they irreversibly differentiated and become cell-committed, and forming germ cells. These EGCs have a developmental capacity which is similar to that of ESCs, although their imprinted expression pattern varies somewhat.

In human, pluripotent ECCs (Andrews et al., 1980; Pera et al., 1989) and EGCs were also established (Shamblott et al., 1998). Human ECCs cells were derived from the undifferentiated stem cell component of germ cell tumors. The human EC clones could be expanded continuously in culture but could also be induced to differentiate and form derivatives from all three germ layers, either in vitro by inducing cell differentiation or in vivo, by teratocarcinoma formation. Compared with human ESCs, human ECC lines are capable of multilineage differentiation in vitro, however, being of tumour origin, these cells are mostly aneuploid, which makes them unsuitable for cell-replacement therapeutics (Andrews et al., 1984). Both human ESC and ECC lines express similar stage-specific cell surface embryonic antigens (SSEAs) and tumour rejection antigens (Tra). Human ECC lines also express Oct-4 (a well characterized pluripotent marker) and form colonies that are morphologically similar to human ESCs. Individual cells in these colonies display a high nuclear to cytoplasmic ratio and visible nucleoli. Several human ECC lines also require the support of a feeder layer to retain pluripotent characteristics, however not all human ECC lines are pluripotent and some feeder-independent human ECC lines have been reported to be nullipotent (i.e. proliferate but have limited capacity to differentiate).

Human EGCs are isolates of PGCs in which are derived from the developing gonadal ridge of 5- to 9-week-old fetuses of elective abortions. Shamblott et al. (1998) reported the successful isolation and characterization of human EGC lines. Human EGCs are pluripotent and are capable of forming all three primordial germ layers. The generally held view is that the pluripotentiality of human EGCs is limited compared with human ESCs, as human EGCs are more advanced in the schema of embryonic development. The number of groups working with human ESCs continues to expand rapidly. This, coupled with the deluge of exciting experimental reports and publications on human ESCs, appears to have overshadowed much of the interest in human EGCs.

Since teratocarcinomas can also be induced by grafting blastocysts to ectopic sites, it was reasoned that it might be possible to derive pluripotent cell lines directly from blastocysts rather than from tumors. This was investigated further in 1981 by Gail Martin and Martin Evans independently in the mouse, and in 1998 by James Thompson in the human. The result was formation of a stable diploid cell line that could generate every tissue of the adult body, including germ cells, and thus named ESCs, which will be explained more latter.

In addition, Jiang et al. in 2002 isolated multipotent adult progenitor cells (MAPCs) from murine bone marrow and demonstrated that these cells expressed telomerase, and that a single MAPC could be expanded clonally to form a large number of identical daughter cells. Under appropriate conditions, MAPCs differentiate into ectoderm, endoderm and mesoderm derivatives and are capable of generating chimeric mice when injected into mouse blastocysts. Reporter genes tagged to MAPCs have been used to identify the contribution of these cells into adult tissues, using a live murine model system. Although the potential of using MAPCs for therapeutic research is high, these cells are particularly rare in the bone marrow and have proven difficult to isolate. It also remains unclear whether these cells are biologically equivalent to human ESCs and if they can be expanded indefinitely while retaining their long term differentiation potential. More data needs to be collected from human MAPCs as most of the current experimental data have been derived from the murine model.

DERIVATION OF ANIMAL EMBRYONIC STEM CELLS

The term embryonic stem cells (ESCs) originates following isolation in 1981 of pluripotent stem cell cultures from the inner cell mass (ICM) of developing mouse blastocysts (Evans and Kaufman, 1981; Martin, 1981). Following, ESCs are most frequently established from the ICM of blastocysts. Mouse ESCs (mESCs) have also been derived from earlier stage embryos such as the 8-cell embryo in the mouse (Delhasie et al., 1996). Furthermore, Eistetter (1989) reported that mESCs cultures might be established from disaggregated morulae with an apparent immediate growth of the colonies from some of the explanted single cells. Up to four separate ESC colonies were founded from a single embryo. Murine

ESCs have also been derived and reported by Brook and Gardner, who used 4.5 day-old hatched, preimplantational mouse embryos flushed from the uterus (Brook and Gardner, 1997). The epiblast was microdissected from the primary endoderm and the trophectoderm and either cultured intact or disassociated into single cells. Both techniques resulted in the formation of ESC cultures, but if the endoderm was left in contact, then no establishment of mESCs were observed. They also reconfirmed the benefit of using delayed blastocysts, showing that the isolated epiblasts of these were the most efficient source of ESC cultures. However, some scientists argue that ESCs are not present in the embryo, but closely resemble the cells of the preimplantation embryo (Brook and Gardner, 1997), in which is not in fact the same (Smith, 2001). The contrasting view held is that embryos (regardless of animal species) contain stem cells. These cells proliferate extensively in the embryo, are capable of differentiating into all the types of cells that occur in the adult, and can be isolated and grown *ex vivo* (outside the organism), whereby they continue to replicate and have the potential to differentiate in vitro and in vivo (Marshak et al., 2001).

ESCs can be stably propagated indefinitely and maintain a normal karyotype without undergoing cell senescence in vitro when cultured in the presence of leukemia inhibitory factor (LIF) (in the case of the mouse) or over a layer of mitotically inactivated mouse embryonic fibroblasts (MEFs), in the monkey and human systems (Figure 1). The frequency of establishing mESC lines depends on individual expertise, the approach chosen and the genetic background. We found that only 2% and 2.9% of mechanical-isolated ICMs derived from the C57BL/6 and BALB/c mouse strains gave rise to ESCs, respectively (Baharvand and Matthaei, 2004). These results are similar to the reports for ESCs derived from the C57BL/6 (Lederman and Burki, 1991; Auerbach et al., 2000), but lower than a report by Kawase et al. (1994), where immunosurgery was used for ICM isolation from trophoectoderm. The frequency of establishing mESC lines with both these strains is substantially lower than the 10–30% frequency reported for the 129 strain (Abbondanzo et al., 1993). The low frequency of ESC production with C57BL/6 and BALB/c strains compared with the 129 strain suggests an undefined genetic difference in the 129, possibly related to the high incidence of spontaneous testicular teratomas and teratocarcinomas observed in this strain (Stevens, 1970). Independent of the aforementioned is that the establishment of mESC lines depends largely on the culture conditions used (Gossler et al., 1986; Hooper et al., 1987; McMahon and Bradley, 1990; Lederman and Burki, 1991). For example, Baharvand and Matthaei (2004) reported that the C57BL/6 ES cell lines could be established when 1000 or 5000 IU/ml LIF was added into the culture medium, whereas 5000 IU/ml LIF was required for the establishment of mESCs when using embryos from the BALB/c strain.

A surprising feature of mouse ESC lines is that the majority of cell lines genetically tested are of male origin (40XY). In female (XX) ESCs, both X chromosomes are active, (which is also observed in the embryonic epiblast) that may result in the unsuitable propagation of ESCs (Rastan and Robertson, 1985). In either case, the XY genotype confers appreciable advantages for germ line transmission. Not only do male chimeras result in significantly higher numbers of offspring than female chimeras, but XY cells can convert the indifferent genital ridge of an XX recipient embryo into testicular development. Because XX germ cells do not develop in a male gonad, this phenomenon of sex conversion results in chimeric males that have spermatocytes which are of ESC origin (Bradley et al., 1984).

In vivo differentiation of pluripotent cells is analyzed (subsequent to transplantation into either syngenic or immunodefficient mice), by the formation of teratocarcinomas or teratomas (Evans and Kaufman, 1981). When ESCs are injected into blastocysts (Bradley et al., 1984) or aggregated with morulae (Wood et al., 1993; Nagy and Rossant, 1993), these cells contribute to all tissues and organs, including the germ line of a chimeric individual and can contribute in the formation of functional gametes (Labosky et al., 1994). Th transmission of genetic manipulated ESCs in vitro can thus be passed into chimeric murine offspring and provide a useful approach for studying varying genetic aspects related to ESCs. Homologous recombination has become a useful transgenic approach for introducing selected mutations into the mouse germ line (Bradley et al., 1984; Koller and Smithies, 1992). These mutant mice are useful animal models for studying gene function in vivo and for clarifying the roles of specific genes in development, metabolic pathways, and immunologic functions.

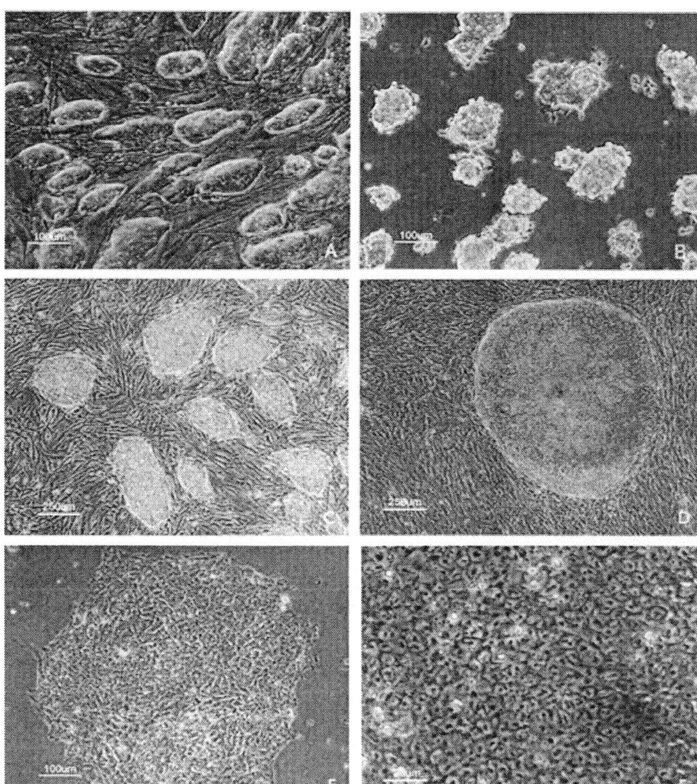

Figure 1. Phase contrast microscopy of (A) mouse embryonic stem cells (Royan B1) on mouse embryonic fibroblasts, (B) mouse embryonic stem cells (ht7) cultured on gelatin-coated dishes in the presence of LIF (a gift from professor Taru Kita, Kyoto university), (C) monkey embryonic stem cells (*Macaca fascicularis*) on mouse embryonic fibroblasts (a gift from professor Taru Kita, Kyoto university), (D) human embryonic stem cells (Royan H5) on mouse embryonic fibroblasts, (E) human embryonic stem cells (Royan H5) on matrigel and in the presence of noggin and bFGF, (F) human embryonic stem cells (Royan H5) on matrigel and in presence of noggin and bFGF with higher magnifcation, transparent nuclei with 1-3 nucleoli.

An important feature of ESCs is their ability to be induced to differentiate in vitro. The differentiation of ESCs can be stimulated by removing feeder cells, differentiation-inhibiting cytokines (e.g., LIF, reviewed in Pedersen, 1994), or by adding specific differentiation inducing substances (e.g., retinoic acid, Rohwedel, 1999). The ability to induce specific and directed cell differentiation in vitro may become a useful mechanism for deriving an unlimited source of pure population cells suitable for transplantation or for the treatment of several diseases. Downing and Battey (2004) have reviewed the various groundbreaking capabilities and technologies that mESCs offer.

ESCs or ESC-likes have been produced in other animal models, including: medakafish from midblastulae stage (Hong et al., 1998), zebrafish from midblastulae stage (Sun et al., 1995), chickens from stage X blastoderm, (Pain et al., 1996), hamsters (Doetschman et al., 1988), rat (Iannaccone et al., 1994; Brenin et al., 1997; Vassilieva et al., 2000; Buehr et al., 2003), mink (Sukoyan et al., 1992), rabbit (Schoonjans et al., 1996), cattle (Strelchenko, 1996; Piedrahita et al., 1998; Cibelli et al., 1998), sheep (Wells et al., 1997), and pigs (Li et al., 2003), however, only mouse and chicken ESCs are capable of colonizing the germ line. In mammals, ES or ES-like cells are derived from the blastocyst stage of pre-implantation development. In non-human primates, ESC lines have been produced in the rhesus monkey (Thomson et al., 1995, Pau and Wolf, 2004), common marmoset (*Callithrix jacchus*, Thomson et al., 1996, *Callithrix jacchus,* Sasaki et al., 2005), and cynomolgus monkey (*Macaca fascicularis*, Suemori et al., 2001). The molecular profile of monkey ESCs has been studied and has been shown to express typical markers of human ESCs and ECCs (Oct-4, SSEA-4, TRA-1–60, TRA-1–81). Monkey ESCs also retain a normal karyotype and have a high capacity to differentiate in vitro and in vivo (Thomson et al., 1995; Kawasaki et al., 2002).

Derivation of Human Embryonic Stem Cells

The first study to describe the successful isolation of human ICM cells and their continued culture for at least two passages in vitro was published by Bongso and co-workers (1994). The authors demonstrated that separated ICM cells differentiate or produce cells with typical hES-cell-like morphology and stained positive for alkaline phosphatase as well as maintained a normal karyotype. Later, in 1998, Thomson and coworkers reported the first human ESC (hESC) lines produced from human blastocyst stage embryos (Thomson et al., 1998). Since then, several studies have reported the establishment of new hESC lines from different sources of embryonic material, including morula (Strelchenko et al., 2004), whole blastocysts, or from isolated ICMs obtained by immunosurgery. Reports of established hESC lines are included in Table-1. In our research 6 hESCs have been derived and fully characterised (Royan H1, Royan H2, Royan H3, Royan H4, Royan H5, and Royan H6) (Baharvand et al., 2004a, Baharvand et al., 2006). Previous reports suggest that the success rate in deriving hESC lines is highly dependent on the quality of recovered blastocysts, isolation conditions used and technical expertise (Pera et al., 2000; Mitalipova et al., 2003).

Comparisons of the success rates for deriving ESCs between research groups is difficult since some of the groups report the numbers of donated embryos, whilst others report the numbers of blastocysts used. In addition, some research groups have used a small number of donated embryos/blastocysts which cannot be statistically analysed. Publication bias may also be of concern, as many groups have been unsuccessful at generating hESC lines and unable to publish their results, or alternately, if they have successfully derived ESCs, their research may not have been novel enough for publication. One common factor in nearly all research describing successful derivation of hESCs is the requirement of embryos to be cultured in the sequential two-step culture system, G1 and G2 media. This biphasic system has been customized for the metabolic and nutrient changes that occur during early human embryonic development and has facilitated in the development of a highly efficient in vitro culture system, which produces developmentally competent blastocysts (Gardner and Lane, 2003; Gardner et al., 2004).

One study reports the use of surplus, low quality or spare embryos which were used for preimplantation genetic diagnosis, for the derivation of hESCs (Pickering et al., 2003, Verlisky et al., 2005, Mateizel et al., 2005, Chen et al., 2005). Another strategy for the derivation of hESC lines was by using an interspecies nuclear transfer (Chen et al., 2003) or cloned blastocyst (created following injection of human somatic cell nuclei into enucleate rabbit oocytes). Recently, one group described the derivation of human blastocyst after heterologous nuclear transfer to donated oocytes (Stojkovic et al., 2005a). This technique is thought to overcome the histocompatible hurdles that current hESC lines endure, as stem cell lines could be produced from individuals with known genetic predisposition to disease.

Surprisingly, it has been demonstrated that even aneuploid zygotes can be used as a source for derivation of hESCs (Suss-Toby et al., 2004). Human ESC lines with genetic disorders have been derived from genetically abnormal embryos diagnosed by preimplantation genetic diagnosis. Verlinsky et al. (2005) reported, the generation of new hESC lines with different genetic abnormalities, which provides an unlimited source of disease cells and an alternate research tool for investigating cellular processes in genetically abnormal cells. The current repository contains 18 hESC lines with genetic disorders, including adrenoleukodystrophy, Duchenne and Becker muscular dystrophy, Fanconi anaemia, complementation group A, fragile-X syndrome, Huntington disease, Marfan syndrome, myotonic dystrophy, neurofibromatosis type I and thalassaemia . In addition, we produced two karyotypically abnormal hESC lines (69 XXY and 69 XXY with chromosomal mosaism) (Baharvand et al., 2006). The establishment of a triploid hESC has also been reported (Heins et al., 2004). Such chromosomal abnormalities are not a surprising feature, since recently (Hardarson et al., 2003) it was discovered that only 42% of surplus ART embryos were chromosomally normal at the blastocyst stage of development.

Human ESCs can be characterized by their immortality, expression of telomerase, pluripotentiality, ability to form teratomas, and maintenance of a stable karyotype even after extended passages in vitro. Expression of Oct-4 and other pluripotent markers e.g., alkaline phosphatase, SSEA -3 and -4, TRA-1-60, TRA-1-81 and GCTM-2 are another key feature of ESCs. For obvious ethical reasons, experiments involving blastocyst injections and ectopic grafting in adult hosts cannot be performed in the human. However, the originally described hESC lines were not clonally derived from a single cell; therefore, their pluripotency could theoretically be attributed to a subset cell population. Human ESCs can be derived, following culture of single-cell clones from established hESC lines. This was accomplished by plating individual cells derived from the parental line under direct observation on individual wells seeded with MEFs. Although the cloning efficiency of the hESCs was relatively poor, a several fold increase was observed when serum-free medium supplemented with basic fibroblast growth factor (bFGF) was used (Amit et al., 2000). The hESC-derived clones retained all the properties of the parental line, including prolonged undifferentiated proliferation with a stable karyotype, expression of high levels of telomerase, and the ability to generate teratomas following in vivo transplantation into immunodeficient mice.

Human and nonhuman primates ESCs share a similar morphology that is distinct from human EGCs and mESCs (Evans and Kaufman, 1981; Thomson et al., 1995; Thomson et al., 1998; Shamblott et al., 1998). Human ESCs form relatively flat, compact colonies that easily dissociate into single cells in trypsin or in Ca^{+2}- and Mg^{+2}-free medium, whereas human EGCs form tight, more spherical colonies that are refractory to standard dissociation methods, and which more closely resemble the morphology of mESC colonies. Moreover, hESCs grow more slowly than mESCs, with a population-doubling time of approximately 12 hours, whereas the population-doubling time of hESCs is about 32-36 hours (Amit et al., 2000, Baharvand et al., 2006). The maintenance of hESCs is not dependent on the addition of LIF for their propagation and for maintenance of pluripotency and this is further reflected by differences in the expression of pluripotent markers. Interestingly, in contrast to mESCs, hESCs with a diploid XX genotype can be readily established and maintained. A lack of X-inactivation and X-chromosome dosage compensation could account for the difficulty in maintaining XX mESCs, but the status of X inactivation in hESCs is unknown. Furthermore, in vitro differentiation of hESCs leads to the expression of alpha-fetoprotein and human chorionic gonadotrophin, which are typically produced by trophoblast cells in the developing human embryo, whereas mESCs are unable to differentiate into this extraembryonic lineage.

One question that should be raised is whether differences between mESCs and hESCs relate to species-specific differences in pluripotency, or reflect differing stages of embryonic development from which the ESCs are derived? Some research indicates the expression pattern of specific antigens is similar in the ICM of the human blastocyst and in hESCs. The pluripotent markers, SSEA-3, SSEA-4 and TRA-1-60 were all found to be expressed in the ICM in a limited series of human pre-implantation blastocysts cultured in vitro (Henderson et al., 2002). As these markers are not expressed in the mouse ICM, this finding suggests that species-specific differences exist in the expression of pluripoten cell markers. Moreover, a general argument in favor of the existence of species-specific differences in pluripotent stem cell phenotype is that pluripotent cells in mice and humans will show differences, irrespective of whether they originate directly from embryos or from primordial germ cells through the protracted process of teratocarcinogenesis (Pera et al., 2000). Thus, at the molecular level, human ECCs resemble hESCs, mouse ECCs resemble mESCs, and the two cell types similar interspecies differences.

Table 1. Published human embryonic stem cell derivation

Reference	Stage of embryo	Establishment Method	Efficiency of establishment	Maintenance		
				Feeder cell	Medium	Passage
Thomson et al., 1998	ICM,	Immunosurgery, using anti BeWo cells	5L of 14B	MEF	DMEM+0.1mM β-ME +1 mM glu+ 1% NEAA+ 20% FCS+ LIF (for initial cultures)	Weekly, 1mg/ml collagenase IV or mechanical slicing
Reubinoff et al., 2000	ICM	Immunosurgery, using anti human whole serum	2L of 4B	MEF	DMEM+0.1M β-mME +2 mM glu+ 1% NEAA+ 20% FCS+ 2000 iu/ml LIF(for initial cultures)	Weekly, 10mg/ml dispase and mechanical slicing
Lanzendorf et al., 2001	ICM	Immunosurgery, using anti BeWo cells	3L of 40B	MEF	DMEM+0.1mM β-ME +1 mM glu+ 1% NEAA+ 20% FCS+ 0.1μl/ml LIF	3-7 days, mechanical slicing
Pickering et al., 2003	ICM	Immunosurgery, using anti human whole serum	3L of 32 B	MEF	DMEM+0.1M β-mME +2 mM glu+ 1% NEAA+ 20% FCS+ (N/A) LIF(for initial cultures)	5-7 days, mechanical slicing
Park et al., 2003		Immunosurgery, using anti human whole serum	3L of 13 isolated ICMs	STO	DMEM+0.1mM β-ME +1 mM glu+ 1% NEAA+ 20% FCS+ (2000U/ml LIF for initial culture and later replaced by) +4ng/ml bFGF	Weekly, mechanical slicing
Hovatta et al., 2003	ICM	Immunosurgery, using anti human whole serum	1L of 5B	Commercial human foreskin fibroblasts (CRL-2429)	Knockout DMEM+0.1mM β-ME +2 mM glu+ 1% NEAA+ 20% FCS+ 1μl/ml LIF	5 days, 2mg/ml dispase and mechanical slicing
Mitalipova et al., 2003	ICM	Immunosurgery, using antiplacental alkaline phosphatase	4L of 19 embryo	MEF	Knockout DMEM+0.1mM β-ME +2 mM glu+ 1% NEAA+ 20% FCS+ 1% ribonucleoside, 1000 U/ml LIF (for initial culture) + 4ng/ml bFGF	N/A
Park et al., 2004	ICM	Immunosurgery, using anti human cell	9 L of 24B	STO	Knockout DMEM+0.1mM β-ME +2 mM glu+ 1% NEAA+ 20% FCS+ 1% ribonucleoside, 2000 U/ml LIF (for initial culture) +4ng/ml bFGF	7 days, 0.04 % collagenase and/or mechanical slicing

Table 1. (Continued)

Reference	Stage of embryo	Establishment Method	Efficiency of establishment	Maintenance		
				Feeder cell	Medium	Passage
Suss-Toby et al., 2004	Whole-embryo culture developed from mononuclear zygote		1L of 9B	MEF	DMEM+0.1mM β-ME +1 mM glu+ 1% NEAA+ 20% FCS+ LIF (for initial cultures)	Weekly 1mg/ml collagenase IV or mechanical slicing
Heins et al., 2004	ICM	Immunosurgery, using anti human whole serum	6L of N/A B	MEF	VitroHES (Vitrolife) +125μg/ml hyaluronic acid + 4ng/ml bFGF	4-5 days, mechanical slicing
Cowan et al., 2004	ICM	Immunosurgery, using anti human red blood cells	17L of 97 isolated ICMs	MEF	Knockout DMEM + 0.055mM β-ME +2 mM GlutaMax+ 1% NEAA+ 10% SR + 10% plasmanate + 5% FCS (for initial culture)+ 5 ng/ml bFGF+ 12ng/ml LIF	0.05% trypsin/0.53 mM EDTA
Baharvand et al., 2004a	ICM	Mechanically	1L of 1B	MEF	Knockout DMEM+0.1mM β-ME +2 mM glu+ 1% NEAA+ 15% FCS+1000iu/ml LIF (for initial cultures)	Weekly, mechanical slicing
Stojkovic et al., 2004	ICM	Immunosurgery, using anti human whole serum	1L of 7B	MEF	Knockout DMEM + 0.1mM β-ME +1 mM glu+ 1% NEAA+ 10% SR + 4 ng/ml bFGF	5-8 days, 1mg/ml collagenase IV or mechanical slicing
Sjogren et al., 2004	ICM or Whole blastocyst culture	Immunosurgery, using anti human whole serum	22L of 114 B	MEF	VitroHES(Vitrolife)	N/A
Oh et al., 2005	ICM or whole-embryo culture	Immunosurgery, using anti human polyvalent immunoglobulins or mechanically	3L of 10B	STO	Knockout DMEM or DMEM/F12+ 0.1mM β-ME +2 mM glu+ 1% NEAA+ 20% SR+ 4ng/ml bFGF	5-7 days, mechanical slicing
Strelchenko et al., 2004	Morula		8L of 46M	MEF or BRL	DMEM or alpha MEM+ 0.1mM β-ME +2 mM glu+ 1% NEAA+ 20% SR or FCS+ 5 ng/ml bFGF	weekly, 1.5 μg/ml collagenase or 1mM EDTA and mechanically
	whole-blastocyst culture		7L of 39B			

Table 1. (Continued)

Reference	Stage of embryo	Establishment Method	Efficiency of establishment	Maintenance		
				Feeder cell	Medium	Passage
Verlinsky et al., 2005	ICM	Immunosurgery, using anti human whole serum	5L of 32ICM			
Lee et al., 2005	ICM	Immunosurgery, using anti human whole serum	3L 0f 7B	human uterine endometrial cells	DMEM/F12+ 0.1mM β-ME +1mM glu+ 1% NEAA+ 20% SR+ 4ng/ml bFGF	7-8 days, mechanical slicing
Genbacev et al., 2005	whole-embryo culture	Mechanically	2L of 111B	Human placental fibroblasts	Knockout DMEM +0.1mM β-ME +1 mM glu+ 1% NEAA+ 20% SR+ 80ng/ml bFGF	5-7 days, 1mg/ml collagenase IV
Inzunza et al., 2005	ICM	Immunosurgery, using anti human whole serum	12L of 37B	Commercial human foreskin fibroblasts (CRL-2429	Knockout DMEM+0.5mM β-ME +2 mM glu+ 1% NEAA+ 20% SR+ 1% ITS	5-8 days, mechanical slicing
Tao et al., 2005	ICM	Immunosurgery, using anti human whole serum	1L 0f 7B	MEF	DMEM+0.1mM β-ME +1 mM glu+ 1% NEAA+ 20% FCS+ 0.1μl/ml LIF	6-7 days, mechanical slicing
Simon et al., 2005	whole-embryo culture	whole-embryo culture	2L of 16B	Human placental fibroblasts	Knockout DMEM +0.1mM β-ME +1 mM glu+ 1% NEAA+ 20% SR+ 12ng/ml bFGF	6-8 days, mechanical slicing
Kim et al., 2005	ICM	Immunosurgery, using anti human whole serum	9 L of 19B	MEF	Knockout DMEM/F12 +0.1mM β-ME +2 mM glu+ 1% NEAA+ 20% SR+ +4ng/ml bFGF	4-7 days, mechanical slicing
Findikli et al., 2005	ICM or whole-embryo culture	Whole embryo culture or immunosurgery, using anti human albumin	9L of 31B	MEF or	Knockout DMEM+0.1mM β-ME +2 mM glu+ 1% NEAA+ 15% FCS+ITS+ 12ng/ml LIF (for initial culture)	Weekly, mechanical slicing
Klimanskaya et al., 2005	ICM	Immunosurgery, using anti human red blood cells	1L of N/A B	Without feeder cells	In sterilized MEF derived ECM+ Knockout DMEM +0.055 mM β-ME +1 mM glutamax+ 1% NEAA+ 8% SR+ 8% plasmanate+ 20 μg/l LIF+16ng/ml bFGF	5-7 days, 0.05% trypsin/0.53 mM EDTA or mechanically or both

Table 1. (Continued)

Reference	Stage of embryo	Establishment Method	Efficiency of establishment	Maintenance		
				Feeder cell	Medium	Passage
Mateizel et al., 2005	ICM	Immunosurgery, using anti human whole serum	5L of 69B	MEF	Knockout DMEM+0.1M β-mME +2 mM glu+ 1% NEAA+ 20% FCS or 20% SR + 1000 iu/ml LIF (for initial cultures) + 4ng/ml bFGF	weekly, 1mg/ml collagenase IV or mechanical slicing
Chen et al., 2005	ICM	Immunosurgery, using anti human whole serum	2L of 117 low quality 3 day-embryos	MEF	Knockout DMEM/F12 +0.1mM β-ME +2 mM glu+ 1% NEAA+ 20% FCS+ +4ng/ml bFGF	3-7d, mechanical slicing in first passages and by 200 IU/ml collagenase
Hong-mei and Gui-an, 2005	ICM	rabbit antiserum to JEG-3 cells	4L of 13 B	MEF	Knockout DMEM/F12 +0.1mM β-ME +2 mM glu+ 1% NEAA+ 20% SR+ +4ng/ml bFGF	7-8d, mechanical slicing in first passages and by 1 mg/ml collagenase or 0.05% typsin/ 0.53 mM EDTA
Wang et al., 2005a	ICM	mannually	3L of N/A	hESC-derived feeder	78% DMEM/F-12+ 0.1 mM β-ME+ 2 mM glu+ 2% NEAA+ 20% SR+ 8 ng/ml bFGF	5-7d, 1mg/ml collagenase IV or mechanical slicing
van de Stolpe et al., 2005	Based on method of Reubinoff et al., 2000					
Baharvand et al., 2006	ICM	Immunosurgery, using anti human whole serum	4L of 5B	MEF	Knockout DMEM+0.1mM β-ME +2 mM glu+ 1% NEAA+ 15% FCS+ITS+1000iu/ml LIF (for initial culture)	Weekly, mechanical slicing

Abbreviations: B: blastocyst, bFGF: basic fibroblast growth factor, β-ME: β-mercaptoethanol, EDTA: ethylene diamintetra acetic acid, ECM: extracellular matrix,glu: L-glutamine, NEAA: nonessential amino acids, SCF: stem cell factor, Flt3L: fetal liver tyrosine kinase-3 ligand, FCS: Fetal calf serum, Knockout DMEM: Knockout Dulbecco's modified Eagle's medium, ITS: insulin-transferin-selenite, ICM: inner cell mass, L: line, LIF: leukemia inhibitory factor, M: morula, MEF: Mouse embryonic fibroblasts, NT-B: nuclear transfer blastocyst, SR: serum replacement. N/A: information not available from published source.

In summary, in order to qualify for establishment of a bona fide ESC line, the following criteria must be satisfied: (i) immortality and telomerase expression; (ii) pluripotentiality and teratoma formation; (iii) maintenance of a stable karyotype after extended in vitro passage; (iv) clonality; ie. a single ESC can give rise to a colony of genetically identical cells or clones, which have the same properties as the original cell; (v) express Oct-4 and other pluripotent markers e.g., alkaline phosphatase, cell-surface glycolipids including: SSEA-1 (in mouse), and SSEA -3 and -4 (in primates), and the cell-surface proteoglycans including: TRA-1-60, TRA-1-81 and GCTM-2, in primate ESCs but not mESCs, and (vi) the ability to contribute to form chimeras. Interestingly, no studies have reported the ability of primate ESCs to participate to form chimeras following blastocyst injection, therefore, the assessment of the developmental capacity of primate ESCs is limited to teratoma formation in xenografts.

MAINTENANCE OF HUMAN EMBRYONIC STEM CELLS

Continuous culture of isolated hESCs in an undifferentiated state whilst maintaining karyotype stability and cell pluripotency is necessary. Both mechanical and enzymatic transfer methods are used to maintain hESC lines (Oh et al., 2005b). The mechanical transfer method is laborious and time-consuming, although remains an efficient technique for the transfer of undifferentiated hESCs and results in similar clump sizes. The enzymatic transfer method is used when the bulk production of cells are required for various experiments. Enzyme treatment results in the more rapid growth and larger production of hESCs. However, the cell clumps vary in size, and there is a higher probability that both differentiated and undifferentiated cells will be transferred. In the case of passaging more differentiated colonies, a combination of both methods allows mass production of hESCs by excluding differentiated colonies from passage by manual selection prior to enzyme treatment. Up until now, no culture methods can effectively allow for the clonal propagation of hESCs.

Another limiting factor relating to cell culture systems is that hESCs still require the presence of feeder layers. Since the first report describing the generation of hESCs by Thomson and coworkers (1998), other investigators have derived and/or maintained additional lines by using MEFs and other cells as feeders (see, Table 1 and Table 2). Transformed MEFs (STO) (Park et al., 2004; Oh et al., 2005a) and buffalo rat liver cells (BRLs) (Strelchenko et al., 2004) have also been used for the same purpose. Human cells have been used for the propagation and derivation of new lines. For example, Richards et al. (2002) produced fibroblasts from three different sources of human cells including fetal muscle (FM), fetal skin (FS), and adult fallopian tube (AFT). Feeders formed from these fibroblasts in combination with culture medium free of animal proteins and supplemented with human serum could support the growth of two hESC lines in an undifferentiated state. However, growth of their hESC lines on FM, FS, and AFT feeders was much slower than those cultured on MEFs. A new hESC line was also derived following culture on human muscle-derived feeders. Hovatta et al. (2003) used commercially available human foreskin fibroblasts for hESC propagation and subsequently derived two new hESC lines by using the same feeders in the presence of fetal calf serum. Furthermore, other types of human cells have been used as feeders for the propagation of hESCs in an undifferentiated state. These include adult bone marrow mesenchymal stem cells (Cheng et al., 2003), adult skin fibroblasts (Richards et al., 2003), foreskin fibroblasts (Amit et al., 2003), adult fallopian tube epithelial cells, human endometrial epithelium (Lee et al., 2005), newborn human foreskin fibroblasts (Amit et al., 2003; Inzunza et al., 2005), commercially obtained human foreskin fibroblasts (Hovatta et al., 2003; Choo et al., 2004), adult skin fibroblasts from abdominal skin biopsies (Richards et al., 2003), human placental fibroblasts (Miyamoto et al., 2004; Genbacev et al., 2005), and fibroblast-like cells derived from spontaneous differentiation of hESCs (Stojkovic et al., 2005c). Establishment of new hESC lines using human feeder cells is necessary to prevent the risks associated with animal feeder cells and for their eventual application for cell replacement therapies.

Furthermore, feeder-free systems for hESC culture is required if hESC cultures are to become clinical-grade, since the use of animal feeders and/or ingredients for growth of hESCs limits the large-scale culture and medical applicability of hESCs. At present, feeder-free systems are not optimal for the derivation and growth of clinical-grade hESC lines since the presence of animal ingredient carries the potential risk for the cross-transfer of different infectious agents. For example, Martin et al. (2005) reported hESC and embryoid bodies could incorporate the nonhuman sialic acid, N-glycolylneuraminic acid (Neu5Gc) from MEFs or from conditioned medium, which resulted in an immune response, mediated by natural antibodies to Neu5Gc which are present in most humans. In effect, these hESCs cells were rejected as foreign cells by the human immune system. Therefore, the identification of growth factors and components within the extracellular matrix (ECM), which allow undifferentiated proliferation of hESCs under feeder-free conditions would be very helpful in the establishment of defined and xenofree culture conditions.

The first attempt to produced feeder-free cultures of hESCs was reported by Xu et al. (2001). They propagated hESCs using Matrigel, an animal-based ECM preparation, or laminin substrates in medium conditioned by MEFs. This system enabled the long-term propagation of the stem cell phenotype, with strong suppression of spontaneous differentiation even at high passages (Carpenter et al., 2004).

Identification of growth factors and components of the ECM, which allow undifferentiated proliferation of hESCs under feeder-free conditions would be very helpful in the establishment of defined and xenofree culture conditions. Lim and Bodnar (2002) analyzed the effects of conditioned medium from STO cell line on hESC culture using two-dimensional electrophoresis mass spectrometry. In total, 136 unique protein species were identified which included some proteins that are known to participate in cell growth and differentiation, extracellular matrix formation and remodeling. Unexpectedly many nominally intracellular proteins were also detected. Some of the identified proteins included insulin-like growth factor binding protein 4, pigment epithelium-derived factor, SPARC (secreted protein, acidic and rich in cysteine), nidogen-2, galectin 1 and peroxiredoxins. This provides further clues as to the proteins that may play a role in cell proliferation and differentiation, however, further studies are required to investigate the hESC proteome and in different feeders, since STOs do not produce sufficient amounts of soluble factors for self-renewal of hESCs and therefore can only be used as feeder cells are not to produce condition medium (Park et al., 2003). Recently, Prowse et al. (2005) identified 102 proteins from conditioned medium of human neonatal fibroblasts which provide invaluable information regarding the factors that may help maintain hESCs. Proteins identified in the ECM (e.g., secreted proteins which are acidic and rich in cysteine, metalloproteinase inhibitors 1 and 2, and the proteoglycans, decorin, lumican, and biglycan) and differentiation and growth factors (e.g., bone morphogenetic protein (BMP), its antagonist gremlin, Wnt inhibitor, Dickkopf homolog 3 and Insulin-like growth factor binding proteins) were considered most likely to be involved in the maintenance of hESC growth, differentiation and pluripotency. These proteins play a role in cell adhesion, cell proliferation and inhibition of cell proliferation. The growth factor, Activin A is paracrinely secreted by MEFs enriches culture medium. Culture medium enriched with recombinant Activin A is capable of supporting the growth of hESCs on laminin coated dishes for >20 passages without the need for feeder layers, conditioned medium from MEFs, or STAT3 activation (Beattie et al., 2005). Human ESCs retained a normal karyotype and expressed markers associated with undifferentiated cells, including Oct-4, Nanog, and TRA-1-60. These hESCs were further characterized as pluripotent following their ability to form teratomas in vivo. A recent report also suggests that Wnt signaling modulation can help to support the growth of hESC cultures short-term and maintain their capacity to express some stem cell markers in the absence of a feeder cell layer (Sato et al., 2004). Based on this idea, a recent study demonstrated noggin, a BMP antagonist, was detected in medium conditioned by human fibroblasts, combined with high bFGF concentrations in medium supplemented with serum replacement which supported the long-term proliferation of undifferentiated hESCs in the absence of feeder cells and or conditioned medium. However, in this case Matrigel coated dishes were used. Unfortunately, application of Matrigel or medium conditioned by mouse embryonic feeder cells is not ideal for potential medical application of hESC because xenogeneic pathogens can be transmitted through culture conditions. (Xu et al., 2005b; Wang et al., 2005b). Another approach was in using human serum as a matrix in combination with medium conditioned by differentiated hESC in an attempt to reduce the exposure of hESCs to animal ingredients, and which is a strategy leading towards the development of completely animal-free conditions (Stojkovic et al., 2005b). In this study, the hESCs grown under these conditions maintained hESC characteristics even after prolonged culture, such as the capacity to differentiate into representative tissues of all three embryonic germ layers both in vitro and in vivo, to self-renew in an undifferentiated state and to maintain a normal karyotype. Genbacev et al. (2005) derived two new hESC lines using serum replacement medium instead of conventional fetal calf serum which were initially cultured on human placental fibroblasts and later transferred to feeder-free conditions using a human placental laminin substrate and defined culture medium (X-VIVO 10 complete medium, proprietary defined medium containing pharmaceutical grade human albumin, recombinant human insulin, and pasteurized human transferrin). One succinct study isolated human ICMs and plated these onto sterilized-ECM (Klimanskaya et al., 2005). The established line was grown in serum-free conditions in human plasmanate and serum replacement medium. To further evaluation the role of growth factor receptors including, bFGF, stem cell factor (SCF), and fetal liver tyrosine kinase-3 ligand (Flt3L), these factors were supplemented into defined medium and hESC growth was monitored (Xu et al., 2005a). Human ESC cultures grown in medium containing bFGF alone or in combination with other factors displayed similar morphological, expression of surface markers and transcription facors to cells grown under MEF-conditioned medium. Similarly these cultures expressed telomerase and were able to differentiation in vitro or in vivo and remained karyotypically stable. In contrast, cells grown in media containing Flt-3L, thrombopoietin, and SCF, individually or in combination with other factors spontaneously differentiated within 6 weeks following culture. Another study reported that the combination of FGF2, TGFβ, LIF and a proprietary serum replacer can achieve serum-free, feeder-free maintenance of hESCs when cultured on a fibronectin ECM (Amit et al., 2004).

The establishment of feeder-free systems for the culture of hESCs is critical for genetic manipulation, i.e. transfection of hESCs or use of small interfering RNAs (siRNAs). Reports indicate that the transfection efficiency of primate ESCs is considerably reduced compared to using mESCs in transfection studies (Furuya et al., 2003). However, hypoxanthine phosphoribosyltransferase-1 and Oct-4 genes have been successfully targeted using homologous recombination (Zwaka and Thomson, 2003) and production of specific siRNAs was effective at downregulating gene expression (Vallier et al., 2004) in hESCs. Homologous recombination could be used as a tool for the repair of specific gene defects in stem cell lines derived from patients suffering disease. It is considered that upon specific induction of stem cell differentiation into the appropriate tissue, these cells could ultimately be transplanted back into the patient (Rhind et al., 2003). This approach coupled with therapeutic cloning offers exciting possibilities for deriving modified and patient-specific hESC lines.

The development of novel culture methods for the propagation of hESCs must undergo stringent testing to ensure that the key properties of pluripotent stem cells are maintained and that the system does not induce karyotypical abnormalities (Draper et al., 2004). For instance, high differentiation rates of hESCs and genomic instability of hESCs after prolonged in vitro growth have been reported (Amit et al., 2003; Carpenter et al., 2004) and some hESC lines have been shown to be partially aneuploid (Carpenter et al., 2004), including the gain of chromosome 17q (Draper et al., 2004), trisomy 20 (Rosler et al., 2004) or aberrantly express the X chromosome (Inzunza et al., 2004). Rosler et al. 2004 reported that pluripotency, the expression of surface markers, and transcription factors remain consistent for up to two years in continuous culture. However, they observed subtle changes in morphology of the cells as well as changes in factors that were identified using microarray technology. Statistically significant differences in SSEA-4 expression, some differences in telomere length, downregulation of collagen, STAT4, a lectin and two genes involved in TGFβ signalling has been described in different hESC lines derived in the same laboratory and cultured under feeder-free conditions (Carpenter et al., 2004; Rosler et al., 2004). This suggests that applied feeders and feeder-free systems for growth of undifferentiated hESC lines are suboptimal since hESC lines show subtle differences in gene expression upon long-term culture (Rosler et al., 2004). In contrast, Brimble et al. (2004) reported hESCs were also maintained on Matrigel or fibronectin-coated plates in MEF conditioned medium, except that EDTA-free trypsin was used to passage the cells. These cultures exhibited multiple definitive characteristics of undifferentiated cells, including long-term self-renewal, expression of markers of pluripotency, maintenance of a normal karyotype, and differentiation to mesoderm, endoderm, and ectoderm.

Table 2. Published culture methods on human embryonic stem cells

Reference	Maintenance		
	Feeder cell	Medium	Passage
Xu et al., 2001	Without feeder	Matrigel coated dish+ conditioned medium of mouse embryonic fibroblasts including Knockout DMEM+0.1M β-ME +1 mM glu+ 1% NEAA+ 20% SR+ 4ng/ml bFGF	Weekly, 200u/ml collagenase IV+ pipetting
Amit et al., 2003	Newborn human foreskin fibroblasts	Knockout DMEM+0.05M β-ME +1 mM glu+ 1% NEAA+ 15% SR+ 4ng/ml bFGF	4-6 days, 1mg/ml collagenase IV
Cheng et al., 2003	Human marrow stromal cells	Knockout DMEM+0.1M β-ME +2 mM glu+ 1% NEAA+ 20% SR+ 4ng/ml bFGF	Weekly, 0.05% trypsin/0.53 mM EDTA followed by the addition of trypsin inhibitors
Richards et al., 2003	Adult skin fibroblasts from abdominal skin biopssies	DMEM+0.1M β-ME +2 mM glu+ 1% NEAA+ 20% FCS	7-8 days, 1.7mg/ml dispase, and mechanical slicing
Xu et al., 2005a	Without feeder	Xu et al., 2001 method with 40ng/ml bFGF or in combination with other factors (15ng/ml SCF or 75 ng/ml Flt3L)	Weekly, 1mg/ml collagenase IV
Genbacev et al., 2005	human placental laminin	in X-VIVO 10 complete medium (proprietary defined medium containing pharmaceutical grade human albumin, recombinant human insulin, and pasteurized human transferring) +0.1M β-ME +1 mM glu+ 1% NEAA+ 80ng/ml bFGF	5-7 days, 1mg/ml collagenase IV
Xu et al., 2005b, and Wang et al., 2005b	Without feeder	On Matrigel coated dishes in DMEM/F12+ 0.1M β-ME+1 mM glu+ 1% NEAA+ 20% SR+ 40 ng/ml bFGF+ 0.5 µg/ml noggin*	2mg/ml dispase
Stojkovic et al., 2005c	autogenic or allogenic fibroblast-like cells derived from the spontaneous differentiation of hESCs	Knockout DMEM+0.1M β-ME+1 mM glu+ 1% NEAA+ 20% SR+ 4ng/ml bFGF	4-5 days, 1mg/ml collagenase IV, or mechanically

Table 2. (Continued)

Reference	Maintenance		
	Feeder cell	Medium	Passage
Klimanskaya et al., 2005	Without feeder cells, on sterilized MEF derived ECM	Knockout DMEM +0.055 mM β-ME +1 mM glutamax+ 1% NEAA+ 8% SR+ 8% plasmanate+ 20 μg/l LIF+16ng/ml bFGF	5-7 days, 0.05% trypsin/0.53 mM EDTA or mechanically or both
Stojkovic et al., 2005b	allogenic fibroblast-like cells derived from the spontaneous differentiation of hESCs	Knockout DMEM+0.1M β-ME+1 mM glu+ 1% NEAA+ 20% SR+ 12ng/ml bFGF+1% of human ITS-A	4-6 days, mechanically
Wang et al., 2005a	hESC-derived feeder	78% DMEM/F-12+ 0.1 mM β-ME+ 2 mM glu+ 2% NEAA+ 20% SR+ 8 ng/ml bFGF	5-7d, 1mg/ml collagenase IV or mechanical slicing
Vallier er al., 2005	Whithout feeder cells, on FCS-coated plates	50% IMDM+ 50% F12 NUT-MIX + 7μg/ml insulin + 15 μg/ml transferrin + 450 μM of monothioglycerol +5mg/ml bovine serum albumin fraction V +10 ng/ml Activin +12 ng/ml FGF2	N/A
Ludwig et al., 2005	Whithout feeder cells, on an ECM composed of collagen IV, fibronectin, laminin and vitronectin	DMEM/F12 base supplemented with human serum albumin, vitamins, antioxidants, trace minerals, specific lipids and cloned bFGF, LiCl, γ-aminobutyric acid (GABA), pipecolic acid and TGF	N/A
Pebay et al., 2005	Whithout feeder cells	DMEM+ 1% ITS+ 0.1 mM β-ME+ 2 mM glu+ 2% NEAA+ 20ng/ml human PDGF+ 10 μM	N/A

*: 0.1 μg/ml noggin is enough, as well. bFGF: basic fibroblast growth factor, β-ME: β-mercaptoethanol, EDTA: ethylene diamintetra acetic acid, ECM: extracellular matrix, glu: L-glutamine, Flt3L: fetal liver tyrosine kinase-3 ligand, FCS: leukemia inhibitory factor, ITS: insulin-transferin-selenite, Knockout DMEM: Knockout Dulbecco's modified Eagle's medium, NEAA: nonessential amino acids, PDGF: platelet derived growth factor, SR: serum replacement, SCF: stem cell factor, S1P: sphingosine-1-phosphate, MEF: Mouse embryonic fibroblasts. N/A: information not available from published source.

Another study has indicated that the use of cell dissociation buffer and enzymatic (collagenase/trypsin) as a means for passaging hESCs, even with limited desegregations of hESCs, could induce chromosomal aneuploidy, especially trisomy 12 and or 17 in conjugation with aberrant quantitative differences in gene expression when compared with using the more time consuming mechanical disassociation technique (Mitalipova et al., 2005). These results confirm previous observations by Buzzard and colleagues (2004) that mechanical means of passaging of hESCs help to retain a stable karyotype, even after 100 passages. Therefore, culture e conditions and passaging are important in the capacity of ESCs for self renewal and maintaining pluripotency.

SELF-RENEWAL AND PLURIPOTENCY

Self-renewal can be defined as the division of a stem cell to generate either one (asymmetric division) or two (symmetric division) daughter stem cells with developmental potentials that are indistinguishable from those of the mother cell (Figure 2). Despite the importance of stem cell self-renewal, mechanisms which govern self-renewal regulation are poorly understood. Studies over the past few years have begun to expand our understanding on self-renewal mechanisms at the molecular level, which have yielded some insights. It has been demonstrated that coordinated events occur via varying signalling pathways for stem cell self-renewal. These include cell signalling pathways conserved between differing types of stem cells as well as pathways which are detected in only specific stem cells. One reason for the requirement of the synergistic activity of multiple pathways may be due to the fact that both proliferation and the maintenance of an undifferentiated state are required for self-renewal. Some pathways that are necessary for self-renewal appear to regulate only proliferation, while other pathways regulate developmental potential and/or differentiation, and some pathways regulate both.

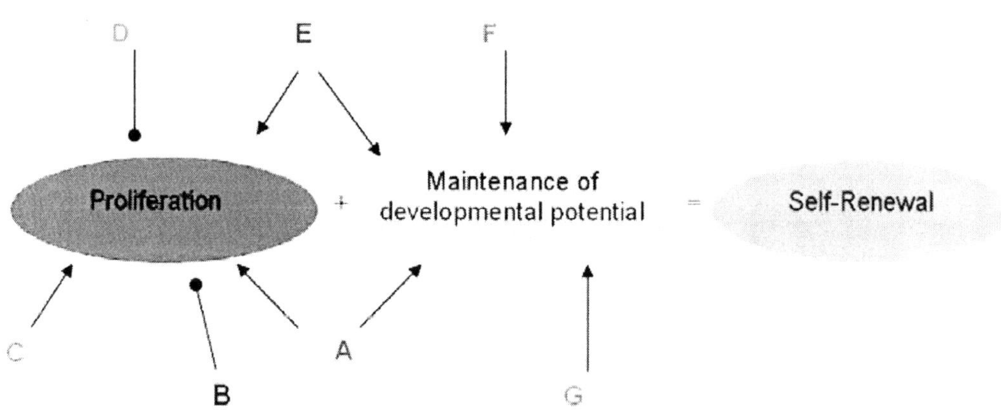

Figure 2. Control of self renewal by regulators with inhibitory or stimulatory effects on the proliferation or maintenance of developmental potential and/or both (adapted from Molofsky et al., 2004, with some modifications).

Stem cells derived from different tissues (adult stem cells) or from embryos obtained (embryonic stem cells) at different stages also appear to have differing mechanisms for self-renewal regulation, as the developmental potential of these cells differ (e.g. pluripotentiality versus multipotentiality) as does their cell fate and constraints on their proliferation (Molofsky et al., 2004).

Embryonic SCs are pluripotent cells that renew themselves in culture and can be propagated indefinitely. ESC self-renewal requires the coordinate action of multiple pathways, some of which are unique to pluripotent cells. Human and mouse ESCs appear to express a set of genes that are found in pluripotent cell populations in the mouse embryo, including some that are known to be important in establishing or maintaining cell pluripotency. Figure 3 illustrates the primary candidate proteins thought to be involved in maintaining cell pluripotency.

Embryonic SCs contain the epiblast/germ cell-restricted POU domain transcription factor, Oct-4 (Scholer et al., 1989; Pesce et al., 1999) and the epiblast-restricted homeodomain protein, Nanog (Chambers et al., 2003; Mitsui et al., 2003). These proteins are necessary for the maintenance of pluripotentiality in the ICM and ESCs (Nichols et al., 1998; Chambers et al., 2003; Mitsui et al., 2003) but not in the regulation pathways in maintaining multipotentiality within adult/somatic stem cells. Some gene products that regulate the self-renewal of ESCs may also regulate the function of somatic stem cells at later stages of development. Stem cells therefore employ different mechanisms to maintain pluripotentiality as compared to multipotentiality, although the mechanisms governing this are unknown.

In ESCs, continuous Oct-4 function at appropriate levels is necessary for maintaining pluripotency. Following less than a twofold increase in Oct-4 expression induces differentiation of ESCs into primitive endoderm and mesoderm, whereas loss of Oct-4 induces the formation of trophectoderm, concomitant with a loss of pluripotency (Niwa et al., 2000). The dosage of Nanog is a critical determinant of cytokine-independent colony formation, and forced expression of this protein confers constitutive self-renewal in ESCs without gp130 stimulation. Nanog may therefore act to restrict the differentiation-inducing potential of Oct-4. Another factor that illustrates the context-dependent nature of self-renewal signals is LIF. This factor binds to its receptor, a heterodimer of the LIF receptor and gp130, which upregulates expression of Stat3 (Niwa et al., 1998). This pathway is directly involved in self-renewal in the mouse, but not in hESCs (Thomson et al., 1998; Daheron et al., 2004; Humphrey et al., 2004). Although *lif* deficiency impairs the self-renewal of mESCs in culture, gastrulation in vivo does not require *lif*, the *lif* receptor, or the gene that encodes gp130 (Dani et al., 1998). Both Nanog and Oct-4 are essential in maintaining ESC identity, but STAT3, following LIF activation, also plays an important role. The expression of gp130 is required for maintaining mouse blastocysts in diapause (Nichols et al., 2001), which is the transient arrest during embryonic development that occurs if mouse embryos are unable to implant in the uterus (usually because the mother is lactating). This suggests that the promotion of ESC self-renewal by LIF may represent a culture adaptation of a mechanism that is employed in vivo to sustain mouse embryos during diapause. When LIF is supplemented into serum-free mESC cultures, it is insufficient in maintaining pluripotency or in blocking (neural) differentiation, which suggests that its activity is dependent upon a second signal. This second signal was recently identified as bone morphogenetic protein (BMP), which induces expression of *Id* ("inhibitor of differentiation") via the Smad pathway. The expression of *Id* results in the blocking of neuronal differentiation via inhibiting the binding of neurogenic helix-loop-helix transcription factors (Ying et al., 2003a). The BMP/Id signaling pathway is a likely candidate, as studies have indicated that the absence of serum in vitro induces ESCs to differentiate into neural cells, under the influence of autocrine FGF (Ying et al., 2003b). In combination with bone morphogenetic protein (BMP), LIF sustains self-renewal, multilineage differentiation, chimera colonization, and germ-line transmission properties. In hESCs, BMPs direct hESC differentiation into trophoblast cells (Pera et al., 2004) or primitive endoderm (Xu et al., 2002b), in conditions that would otherwise support their undifferentiated proliferation. Similarly, Beattie et al. (2005) demonstrated that BMP-4 is ineffective in maintaining pluripotency in hESCs. Recently, Xu et al. (2005b) demonstrated BMP2 and BMP4 proteins were detected at an increased level in unconditioned medium-cultured hESCs compared with cells cultured in either conditioned medium or in unconditioned medium containing noggin and bFGF. Both human *ID1* (Lopez-Rovira et al., 2002) and mouse *Id1* (Korchynskyi et al., 2002) promoters contain BMP-responsive elements, and have been shown to be targets of BMP4 signaling in both human (Xu et al., 2005b) and mouse (Ying et al., 2003a) ESCs, respectively. They therefore examined the expression of *ID* genes (as a second indicator of BMP signaling activity) in hESCs cultured in various media. Transcripts positive for *ID1−4* were more abundant in hESCs cultured for 24 h in unconditioned medium or conditioned medium plus BMP4 than in cells cultured in conditioned medium. The addition of noggin to unconditioned medium significantly reduced *ID gene* expression. This demonstrated that factors blocking lineage-specific differentiation could be critical for the mechanism of self-renewal, suggesting that different combinations of such factors may be required by different stem cells, or stem cells in different environments. Blockade of lineage-specific transcription factors by Id proteins enables the self-renewal response to LIF/STAT3 signaling (Ying et al., 2003a). MEK/ERK signaling is also involved in ESC self-renewal and differentiation. Inhibition of MEK/ERK by the MEK inhibitor PD098059 inhibits differentiation and maintains ESC self-renewal in culture, and the expression of ERK and SHP-2 is thought to counteract the proliferative effects of STAT3 and promote differentiation (Burdon et al., 1999, Burdon et al., 2002). The interactions between Nanog, Oct-4, LIF signaling and BMP signaling in the maintenance of mESCs have not been defined and are most likely complex interactions that may be context dependent (Ying et al., 2003a). Nanog overexpression can partly bypass the requirement for LIF and BMP stimulation, but Nanog does not appear to be transcriptionally downstream of LIF signaling, and differences in the phenotypes of Nanog-deficient and Stat3-deficient mESCs suggest that these pathways are at least partially independent (Chambers et al., 2003; Mitsui et al., 2003; Ying et al., 2003a).

A more recent study has implicated Wnt-signaling pathways in the maintenance of ESC pluripotency. The Wnt pathway is activated by a specific pharmacological inhibitor (6-bromoindirubin-3-oxime) of glycogen synthase kinase-3 (GSK-3) which maintains the undifferentiated phenotype in both mouse and hESCs and sustains expression of the pluripotent stage-specific transcription factors, Oct-4 and Nanog (Sato et al., 2004). The reversibility of the BIO-mediated Wnt-activation in hESCs also suggests a practical application of GSK-3-specific inhibitors in regulating events during early differentiation, which may prove valuable for the derivation of specific cell types suitable for regenerative medicine.

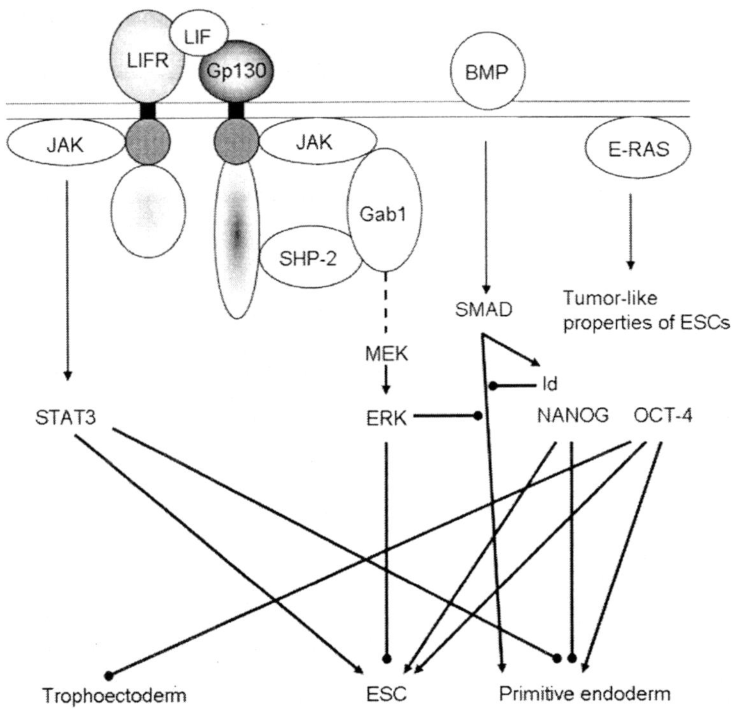

Figure 3. Regulation of self-renewal in mESC. The LIF receptor complex contains a LIF-specific chain (LIFR) and gp130 (Heinrich et al., 1998). Ligand binding induces cross-phosphorylation of Jak kinases, phosphorylation of receptors cytosolic Tyr residues, which function as docking sites for SH2-containing molecules, like STAT3 and the SHPTP2 phosphatase (Watowich et al., 1996; Heinrich et al., 1998). Recruited STAT3 becomes phosphorylated, dimerizes, and is translocated to the nucleus where it stimulates expression of genes implicated in cell renewal. SHPTP2 connects the LIFR complex to MAP kinase activation. The MEK/ERK signaling pathway prevents ESC self-renewal (Burdon et al., 1999). In parallel, intrinsic ES proteins, such as Oct4 and Nanog act in concert to induce gene expression to promote ESC renewal and pluripotency (Mitsui et al., 2003; Chambers et al., 2003). Nanog and STAT3 may have overlapping targets but act independently. The LIF and BMP factors synergistically block differentiation. The LIF factor activates STAT3 and blocks non-neural differentiation, whilst BMP blocks neural differentiation by induction of Id's. Embryonic stem cells contain a constitutively active Ras protein (E-Ras), which is responsible for the in vivo tumor-like properties of injected ES cells (Takahashi et al., 2003). In contrast, over-expression of Oct4 results in differentiation of ES cells into primitive endoderm; whereas over-expression of Nanog leads to a LIF-independent ES cell self-renewal. The growth factor, Oct4 and Nanog signaling prevents differentiation into trophectoderm and primitive endoderm, respectively.

Multiple coordinating pathways and the combination of intrinsic and extrinsic signals are required for the stoichiometric balance of cell self-commitment to either self-renew or differentiate. Following an imbalance in any single cell can result in loss of ESC identity. Other candidate genes for maintaining cell pluripotentiality include, GBX2 (Chapman et al., 1997), UTF1 (Okuda et al., 1998), Pem (Fan et al., 1999), L17 (Rodda et al., 2001), Rex-1, FGF-4, Cripto, and FOXD3. The acidic zinc finger transcription factor, Rex-1, has been shown to be a specific regulatory target of Oct-4 in mESCs (Rogers et al., 1991; Hosler et al., 1993; Ben-Shushan et al., 1998). Similarly, the fibroblast growth factor 4 (FGF-4) expressed within the ICM (Rappolee et al., 1994) is another transcriptional regulatory target of Oct-4 in ESCs (Yuan et al., 1995). The homeobox domain transcription factor, SOX2 (Avilion et al., 2003) and Cripto/TDGF1 are expressed in the ICM and trophoblast cells of the mouse blastocyst, in mESCs, and in human and mouse ECCs (reviewed in Adamson et al., 2002). The genes FOXD3 or GENESIS are expressed during early mouse embryonic development and ESCs, and following abrogation of FOXD3 function, embryos fail to develop beyond the four-cell stage (Sutton et al., 1996; Hanna et al., 2002; Carpenter et al., 2003). In addition, FOXD3 has been found to be critical for endodermal differentiation, and which antagonizes the activity of Oct-4 (Guo et al., 2002). These genes are all expressed in the ICM of blastocysts and are downregulated upon differentiation; however, they are not exclusively expressed in pluripotent embryonic stem cells, but are also expressed in other cell types in the soma. Their precise roles and potentials for maintaining pluripotentiality or self-renewal remain to be determined.

Identification of the molecular signature of undifferentiated hESCs and their differentiated progeny including ligand/receptor pairs and secreted inhibitors of different signaling pathways including fibroblast growth factor (FGF) and/or TGFβ/bone morphogenetic protein (BMP) pathways (Sato et al., 2003). The presences of these factors are essential for in vitro culture conditions, as they prevent differentiation but also induce differentiation of hESCs. The use of these factors in medium is extremely difficult, since specific growth factors have individual and unique effects on the growth of undifferentiated hESCs and their differentiation (Schuldiner et al., 2001). For instance, one of the most routinely used mitogenic and growth factor for propagation of undifferentiated hESCs is bFGF which can induced cells to form ectodermal and mesodermal cells (Schuldiner et al., 2000) and can also act as a differentiation factor to induce the formation of immortalized hypothalamic neurons (Park et al., 2004).

Numerous techniques have been used for characterizing the gene expression profile of stem cells which is most often achieved using a comparative approach to compare ESCs with their differentiated counterparts or with somatic or nonpluripotent cell types. These techniques include serial analysis of gene expression (Tremain et al., 2001; Zhou et al., 2001; Anisimov et al., 2002; Richards et al., 2004), subtractive hybridization (Graf and Torok-Storb, 1995), representational difference analysis (Geschwind et al., 2001), cDNA microarray (Aprelikova et al., 2001; Ramalho-Santos et al., 2002; Ivanova et al., 2002; Kriz et al., 2003; Fortunel et al., 2003; D'Amour and Gage, 2003; Sato et al., 2003; Abeyta et al., 2004; Bhattacharya et al., 2004), and oligonucleotide microarray (Kelly and Rizzino, 2000; Phillips et al., 2000; Loring et al., 2001; Terskikh et al., 2001; Park et al., 2002; Tanaka et al., 2002; Sperger et al., 2003; Easterday et al., 2003; Luo et al., 2003; Cai et al., 2004) technologies. These methodologies have also been used to decipher genes that are specifically expressed in, ESCs, mesenchymal stem cells, neural stem cells (NSCs), and/or hematopoietic stem cells (HSC). Although each technique has its own unique advantage, cDNA and oligonucleotide microarray technologies have gained precedence as they provide opportunities for comparison of global gene expression patterns between different populations. Specifically, they are proving to be invaluable tools in elucidating the molecular mechanisms involved in the maintenance of pluripotency and initial differentiation events of hESCs. For example, a global transcriptional profile to identify the state of 'stemness' (the molecular makeup of stem cells) in various hESC lines (Sato et al., 2003). A set of 918 genes enriched in undifferentiated hESCs was determined following comparison with their differentiated counterparts. Some of these genes included ligand/receptor pairs and secreted inhibitors of the FGF, TGFβ/BMP, and Wnt pathways, highlighting a prevalent role for these pathways in hESCs. Importantly, a significant number of hESC-enriched genes, including several signaling components intersected those previously published from mouse ESC data, indicating that a 'core molecular programme' is shared between stem cells from the two species. More recently, serial analysis of gene expression (SAGE), which is a more quantitative approach, has been used to obtain the transcriptome profiles of two hESC lines which were cross compared with both murine ESCs and other human tissues (Richards et al., 2004). This study further supported the notion that human and mouse ESC appear to share a number of expressed gene products. However, numerous notable differences were also reported, including an inactive LIF pathway and the high preponderance of several important genes, such as POU5F1 (Oct-4) and SOX2, in hESCs. This group also established a list of genes comprised of known ES-specific genes and new candidates that may serve as novel hESC markers and which might contribute towards defining the 'stemness' phenotype. Although these studies provide insight into different pathways involved in hESC pluripotency, more profound in vitro studies are needed to map the self-renewal and pluripotency mechanisms of ESCs.

Differentiation of Embryonic Stem Cells In Vitro

Upon removal of ESCs from feeder layers and subsequent transfer to suspension cultures, ESCs begin to differentiate into 3-dimensional, multicellular aggregates, forming both differentiated and undifferentiated cells, termed embryoid bodies (EBs). Initiation of differentiation may also be induced following addition of cells into two-dimensional cultures, i.e. on a differentiation inducing layer such as a matrix or feeder cells. The formation of EBs can be acquired by culturing ESC aggregates in suspension (i.e. using a non-adherent dish), or within hanging-drops followed by cultivation in Petri dishes. Embryoid bodies are composed of a haphazard collection of precursor and more fully differentiated cells from a wide variety of lineages and resemble early post-implantation embryos (Shamblott et al., 2001). Following prolonged culture, they eventually forming cavities and cystic embryoid bodies (Itskovitz-Eldor et al., 2000). In vitro differentiation experiments have indicated that plated cultures of embryoid bodies display a variety of different morphologic types, including rhythmically contracting cardiomyocytes, pigmented and nonpigmented epithelial cells, neural cells with outgrowths of axons and dendrites, and mesenchymal cells (Odorico et al., 2001). Also, hESCs-derived embryoid bodies express genes specific for each of the three embryonic germ layers, including α-fetoprotein, neurofilament 68-kd subunit, β- globin, and α–cardiac actin marking primitive endoderm, neuroectoderm, and mesoderm derivatives (Itskovitz-Eldor et al., 2000; Schuldiner et al., 2000). Recent studies have also demonstrated ESC differentiation into germ cells and more mature gametes, although significant unanswered questions remain about the functionality of these cells (Hubner et al., 2003; Geijsen et al., 2004; Clark et al., 2004; Toyooka et al., 2003; Lacham-Kaplan 2004; West and Daley, 2004). The derivation of germ cells from embryonic stem cells in vitro provides an invaluable assay both for the genetic dissection of germ cell development and for epigenetic reprogramming, and may one day facilitate nuclear transfer technology and infertility treatments.

The precise inductive elements that regulate embryonic pattern formation are still unclear. However, these cell formations are a useful tool for investigating methods relating to extraction of specific cell types from mixed cell populations or heterogeneous teratomas and to perhaps study the differentiation events of precursor cells towards a particular cell lineage. Feasible methods that may help to achieve this include; (i) the addition of specific combinations of growth factors or chemical morphogens, (ii) co-culture or transplantation of ESCs with inducer tissues or cells, (iii) implantation of ESCs into specific organs or regions, (iv) over expression of transcription factors associated with development of specific tissues, (v) selection of cells that activate a particular lineage-specific program of gene expression, and finally (vi) isolation of cells of interest by gene tagging and fluorescence-activated cell sorting (FACS) (Li et al., 1998a; Schuldiner et al., 2000; Soria et al., 2000; Weissman, 2000; Odorico et al., 2001). Some of these methods have already been applied for the enrichment of ESC cultures for a particular cell type in vitro. However, little research has been conducted in vivo (Odorico et al., 2001). Recent reports on the differentiation of ESCs into neurons, cardiomyocytes and insulin-secreting cells demonstrate the ability of ESCs to form cells representative of the three germ layers that are ectodermal, mesodermal and endodermal in nature, respectively.

Differentiation of Embryonic Stem Cells into Neurons

The development of the mammalian nervous system is a complex event that is controlled by a number of epigenetic and endogenous cellular signals during embryogenesis and early postnatal life (McKay, 1989; Luskin, 1994). The signals that control the specific differentiation events are not well understood, due to the difficulty in dissecting and analyzing the innate and interactive processes of neuronal cell differentiation and migration during embryogenesis. Primary cultures and embryonic precursor cell lines like PC12, SHSY- 5Y (Greene and Tischler, 1976; Pfeiffer et al., 1981) have been used as models to study these processes. However, cells lines are not ideal for studying cell commitment into neuroectoderm. The use of ESC-differentiation systems have been suggested as an alternative model for studying these events. The list of neurological diseases that may be treatable using differentiated human ESCs is vast. If the in vitro differentiation process of ESCs could be controlled to form neuroectoderm cell lineages in vivo, then it is anticipated that intraepithelial precursor cells could be generated and further differentiated into both neuronal and glial cell lineages (Rathjen et al., 2002). The first evidence for the capacity of ESCs to differentiation in vitro into the neuronal lineage was reported by Wobus et al., (1988, 1994) and Rohwedel et al., (1994). Retinoic acid (RA) (Bain et al., 1995; Fraichard et al., 1995; Schuldiner et al., 2001; Wichterle et al., 2002), noggin (Gratsch and O'Shea, 2002) and growth factors such as bFGF and epidermal growth factor (EGF) (Okabe et al., 1996; Reubinoff et al., 2001; Zhang et al., 2001) have also been used to induce neuronal differentiation from ESCs.

Based on earlier findings by Okabe et al. (1996) efficient protocols for growth factor-mediated lineage selection of neuronal cells were established (Okabe et al., 1996; Lee et al., 2000; Baharvand, et al., 2004b). These protocols include: (i) formation of cells of all three primary germ layers in EBs; (ii) selective differentiation of cell populations expressing nestin, an intermediate filament protein normally found in neural precursor cells (Lendahl et al., 1990) by fetal calf serum depletion. In serum-free conditions and following the addition of insulin-transferine-selenium-fibronectin (ITSFn), many cell types die which selects for a proportion of nestin-positive cells (Lee et al., 2000) (iii) proliferation and maintenance of neural precursor cells in the presence of inductive factors such as basic fibroblast growth factor (bFGF), epidermal growth factor (EGF) (Okabe et al., 1996), FGF8, and Shh (Lee et al., 2000); and (iv) the differentiation induction and maintenance of functional neurons and glial cells by the combined addition of neural differentiation and survival-promoting factors.

Two papers published in 2001 (Reubinoff et al., 2001; Zhang et al., 2001) highlighted how the potential of hESC could be harnessed towards the generation of brain cells. Zhang et al. (2001) have combined these techniques which were initially developed for the neural differentiation of mESCs (Okabe et al., 1996; Lee et al., 2000) and adapted these to produce neural stem cells. This occurs via a successive stepwise approach, which consists of inducing the formation of EBs and from these generating neural rosettes, which are proliferating structures that mimic neural tube formation. Rosettes are subsequently harvested by selective dissociation and are cultured as free-floating aggregates of neural precursors, capable of generating neurons and glia. Reubinoff et al. (2001) chose an alternate and simpler method to produce similar results. Neural differentiation was induced by overgrowth of undifferentiated ESCs (Reubinoff et al., 2000). Maintaining hESCs in culture without passage or replenishing feeder cells led to spontaneous neural differentiation within a heterogeneous population of ESC progeny. Individual clusters of presumptive neural progenitors were identified by phase-contrast microscopy and manually transferred onto uncoated culture dishes. Following culture in defined medium supplemented with bFGF and epidermal growth factor (EGF), these cells formed aggregates highly enriched with neural precursor cells (Figure 4).

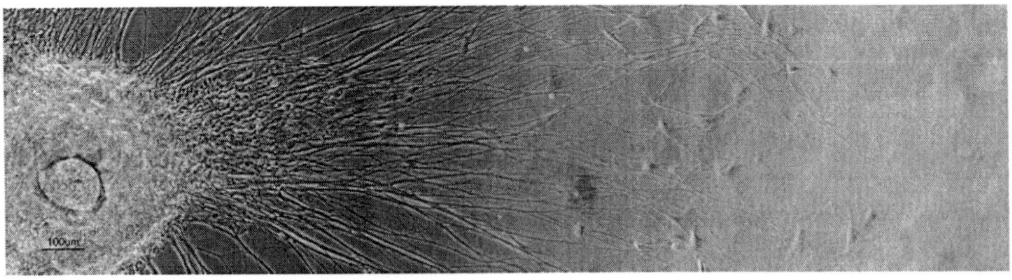

Figure 4. Differentiation of neuronal cells from human embryonic stem cells (Royan H5). The aggregation of cell bodies is displayed on the far left, with cell projections extending to the right.

After withdrawal of bFGF and EGF, downregulation of nestin and mash-1 is followed by upregulated expression of neuron-specific NFM, synaptophysin, Nurr1, and tyrosine hydroxylase (TH) genes. A decreased formation of nestin-positive cells is assimilated with an increased number of neuronal cells expressing neuron-specific proteins. Mature neuronal cells are evidenced by the production of neurotransmitters such as dopamine, serotonin, GABA, and glutamate. These results suggest that in the presence of neuronal differentiation factors, such as retinoic acid, FGF4, FGF8, or bFGF, ESC-derived cells, led to the enrichment of cholinergic, serotinergic, dopaminergic and GABAergic neurons, respectively, (Okabe et al., 1996; Lee et al., 2000; Rolletschek et al., 2001, Barberi et al., 2003). Neuronal differentiation was induced by removal of bFGF from culture and following the addition of ascorbic acid (Shin et al., 2004).

Previous studies using primary cultures of dopaminergic cells have shown that interleukin-1 (IL-1; Ling et al., 1998), dibutyryl-cAMP (Branton et al., 1998), transforming growth factor-3 (TGF3; Krieglstein et al. 1998), glial cell line-derived neurotrophic factor (GDNF; Beck et al., 1995; Eggert et al., 1999), and neurturin (NTN; Horger et al. 1998) could promote the differentiation and survival of dopaminergic neurons. Similarly, TH- and dopamine transporter (DAT)-positive neurons were significantly increased, although no increases on the proportion of GABA- and serotonin-positive neurons were reported (Rolletschek et al., 2001). Recently, Takagi et al. (2005) generated neurospheres composed of neural progenitors from monkey ESCs, which are capable of producing large numbers of dopaminergic neurons. They demonstrated that FGF20, which is preferentially expressed in the substantial nigra, acts synergistically with FGF2 to increase the number of dopaminergic neurons in ESC–derived neurospheres. They also analyzed the effects following the transplantation of dopaminergic neurons generated from monkey ESCs into 1-methyl-4-phenyl-1,2,3,6-tetrahydropyridine–treated (MPTP-treated) monkeys (a primate model for Parkinson disease). Behavioral studies and functional imaging revealed that the transplanted cells could attenuate MPTP-induced neurological symptoms. Furthermore, MS5-feeder cocultured with mESCs, induced differentiation of ESCs into midbrain dopamine neurons (Barberi et al., 2003). The discovery of a strong neuralizing activity, termed stromal cell–derived inducing activity (SDIA) is located on the surface of stromal cells (Kawasaki et al., 2000) has been used to derive neural precursor cells from ESC cultures. In the absence of exogenous bone morphogenic protein–4, mESCs differentiate efficiently into neural precursors and neurons when cultured for 1 week on SDIA-expressing mouse stromal cells (PA6 cells) (Kawasaki et al., 2000). Recently, SDIA induction has also been applied to primate ESCs, resulting in the generation of large numbers of neural precursors and postmitotic neurons when cultured on PA6 cells for two weeks (Kawasaki et al., 2002). Also, Barberi and colleagues (2003) were able to induce neuronal differentiation by coculturing mESCs with murine bone marrow-derived stromal feeder cell lines, MS5 and S17, or with primary stromal feeder cells obtained from aorta-gonad mesonephrous region.

Li et al. (2005) differentiated hESCs into spinal motoneurons using retinoic acid and in the presence of sonic hedgehog. The in vitro-generated motoneurons expressed HB9, HoxC8, choline acetyltransferase and vesicular acetylcholine transporters, which induced a clustering of acetiylcholine receptors in myotubes that were shown to be electrophysiologically active. In addition, motor neurons generated using this technique was able to integrate into host tissue following cell transplantation experiments into mice (Wichterle et al., 2002).

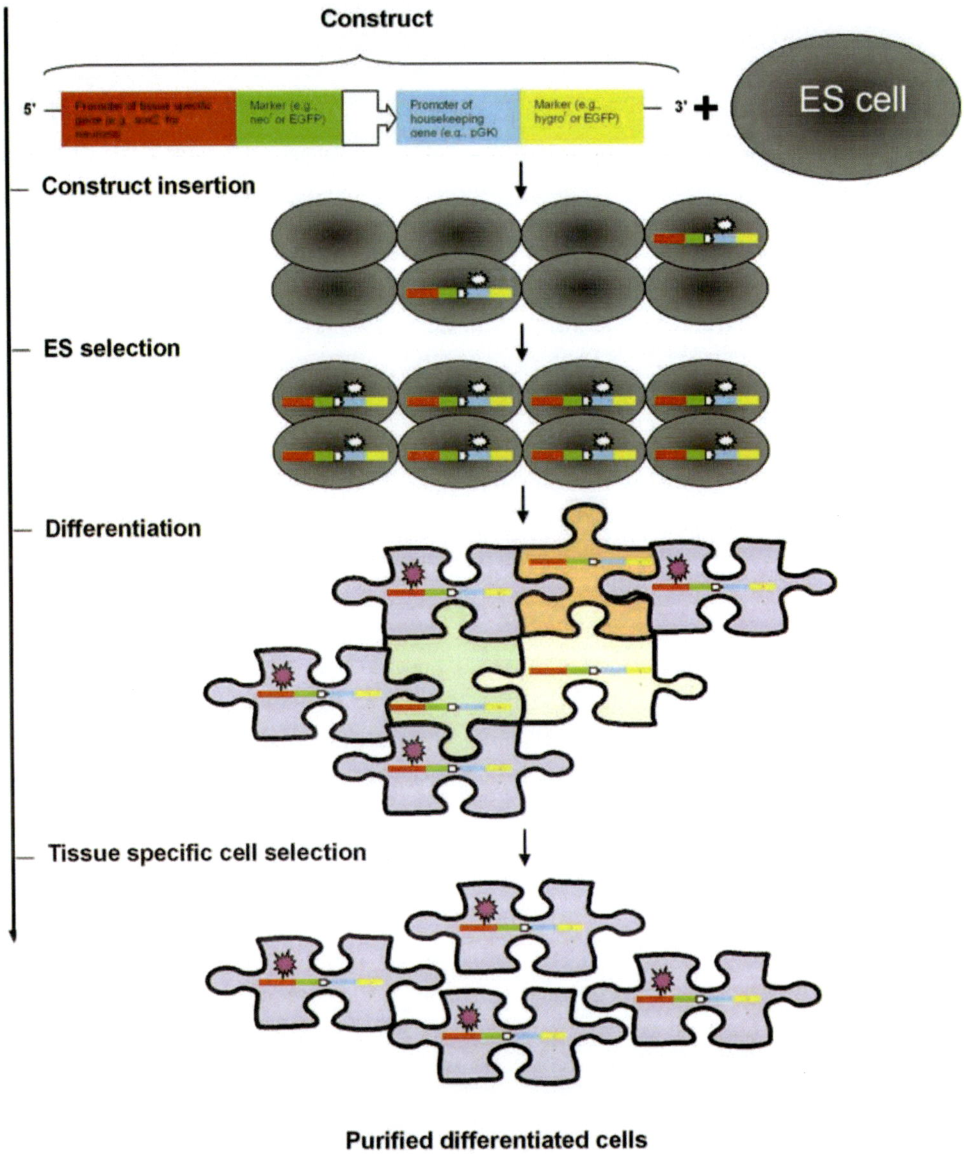

Figure 5. Structure of a tissue specific promoter (gene of interest /housekeeping gene) and methods for selection and purification of transgene following transfection into ESC-derived differentiated cells.

Recently, Watanabe et al. (2005) demonstrated directed differentiation of telencephalic precursors from mESCs using an optimized serum-free suspension culture system (SFEB culture). Treatment with Wnt and Nodal antagonists (Dkk1 and LeftyA) during the first 5d of SFEB culture resulted in almost pure (~ 90%) selective neural differentiation in ESCs. In the presence of Dkk1, with or without LeftyA, SFEB induces efficient generation (~35%) of cells expressing the telencephalic marker Bf1. Addition of Wnt3a during the late culture period increases the pallial telencephalic population ($Pax6^+$ cells yield up to 75% of $Bf1^+$ cells), whereas sonic hedgehog has been found to promote basal telencephalic differentiation (into $Nkx2.1^+$ and/or $Islet1/2^+$ cells) at the cost of pallial telencephalic differentiation.

Transgenic approaches have also been used for the selective isolation of neuronal cells from ESC lines via use of tissue-specific promoters, fused to selectable marker genes (Figure 5). For example, neural precursors were selectively generated from ESCs transfected with the SRY-related transcription factor, Sox2 (Li et al., 1998b). Following addition of RA, G418 was applied to select for Sox-2 positive, neural cells from differentiating ESC cultures. Almost all of the cells expressed Sox2 and Sox1, consistent with the expression of nestin at early stages. Selected Sox2 positive precursor cells were able to further differentiate and form mature neurons, evidenced by GABA and glutamate neurotransmitters (Li et al., 1998b).

Bibel et al. (2004) reported that selection of highly proliferative mESCs expressing green fluorescent protein (GFP) from the *Mapt* (*tau*) locus, followed by treatment with retinoic acid, generated an almost pure population of precursors which were Pax-6-positive, which is a specific marker for radial glial cells. Mimicking *in vivo activity*, these cells formed neurons and displayed remarkably uniform biochemical and electrophysiological characteristics.

Differentiation of Embryonic Stem Cells Into Cardiomyocytes

Primary myocardial cell lines are useful models for investigating ultrastructural and functional features related to terminal differentiation in the heart. Inherent problems associated with breeding, collection and preparation of early mammalian embryos has meant few investigators have utilized embryonic tissues for ultrastructural studies of mammalian myocardium. In addition, cell isolation techniques used to obtain these primary cultures are not suitable for the study of prenatal early cardiac development, due to the small quantity of myocardial material retrieved and the difficulty in maintaining cells is an immature state during in vitro culture. Therefore, several attempts have been made to generate permanent cell lines from embryonic rat heart (Kimes and Brandt, 1976), embryonic avian heart (Jaffredo et al., 1991), adult rat myocardium (Caviedes et al., 1993), myocardial tumor tissue of transgenic mouse (Field, 1993), and proliferating cardiac cells by virus transfection technology (Sen et al., 1988). Disadvantages in using adult derived cell lines have hampered progress in developing clinical grade lines for cell therapeutics, due to their functional, structural and molecular limitations. Furthermore, these cell lines may not be appropriate for cell therapies, as molecular analysis of these lines indicate only a few markers specific for differentiated heart cells are expressed. In addition, the absence of key myocardial features, which relate to cell functionality, include lack of cell contraction and action potential generation. Propogation of ESCs and directed development into heart specific cell types, may serve as a better model for studying early events relating to mammalian heart development and for studying repair mechanisms in myocardium.

Contractile or beating cardiomyocytes arise within mESC and hESC cultures following spontaneous differentiation (Wobus et al., 1991; Maltsev et al., 1999; Itskovitz-Eldor et al., 2000; Baharvand et al., 2004a; Baharvand et al., 2005) (Figure 6). Cardiomyctes derived from ESCs are easily detected in heterogenous cell populations such as EBs by their spontaneous beating activity. These spontaneously contracting areas appeared in 8-10% and more than 80% of the hESCs and mESCs-derived EBs, respectively and displayed structural and functional properties of early-stage cardiomyocytes (Wobus et al., 1991; Kehat et al.,

2001). Similar to in vivo cardiogenesis (Manasek, 1970; Chacko, 1976), previous studies have shown that during in vitro ES differentiation into cardiomyocytes, myofibrils were initially distributed in sparse, irregular myofibrillar arrays, which gradually matured into parallel arrays of myofibrils and ultimately aligned into densely packed sarcomeres. We found that the Z-line, I- and A-bands, and intercalated discs including adherence junctions, gap junctions and desmosomes were distinct in ES-derived cardiomyocytes. The presence of T-tubules, H-band, and M-band could also be distinguished in some cells (Figure 7) (Baharvand et al., 2005).

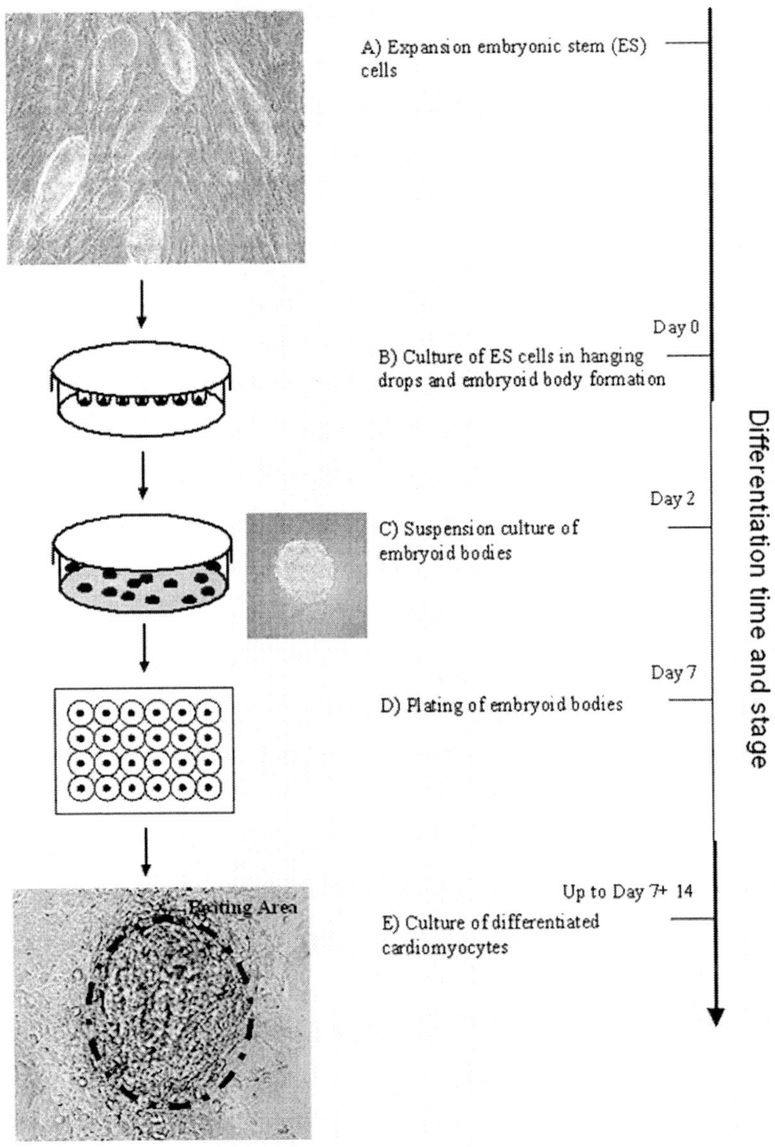

Figure 6. Typical method used to induce mESC (Royan B1) differentiation into cardiomyocytes in vitro, spontaneously.

Hescheler et al. (1997) and Kehat et al. (2001) reported that compared with adult cardiomyocytes, T-tubules, H-band, and M-band were missing in mES and hESC-derived cardiomyocytes. The T-tubular system and M-band are thought to form at the terminal end of myofibrillar maturation (Viragh and Challice, 1969). Our research has shown that the T-tubule and H-band, but not the M-band are visible during long-term culture (40 days) within human-ESC-derived cardiomyocytes (Baharvand et al., 2004a). It has also been reported that M-bands in a small proportion of cardiomyocytes could be detected at 60 days following plating of EBs derived from hESCs. Interestingly, T-tubules were not observed in any of the examined cells (Snir et al., 2003), but were detected in cardiomyocytes cultured on cardiogel (a naturally occurring ECM derived from cardiac fibroblasts), which contains a complex mixture of laminin and fibronectin.

Cells within the EB differentiate into a wide range of cardiomyocytic cell types, including atrial, ventricular, Purkinje, sinus nodal and pacemaker-like cells. These different phenotypes are also associated with the expression of cardiac-specific marker genes, such as ion channels, structural proteins and receptors (Wobus et al., 1991; Hescheler et al., 1997) and can be identified by their typical action potentials (Hescheler et al., 1997) or pharmacological properties. Microelectrode arrays (MEA) have been used to detect key features such as a functional syncytium with stable focal activation and conduction properties in hES-derived cardiomyocytes (Kehat et al., 2002).

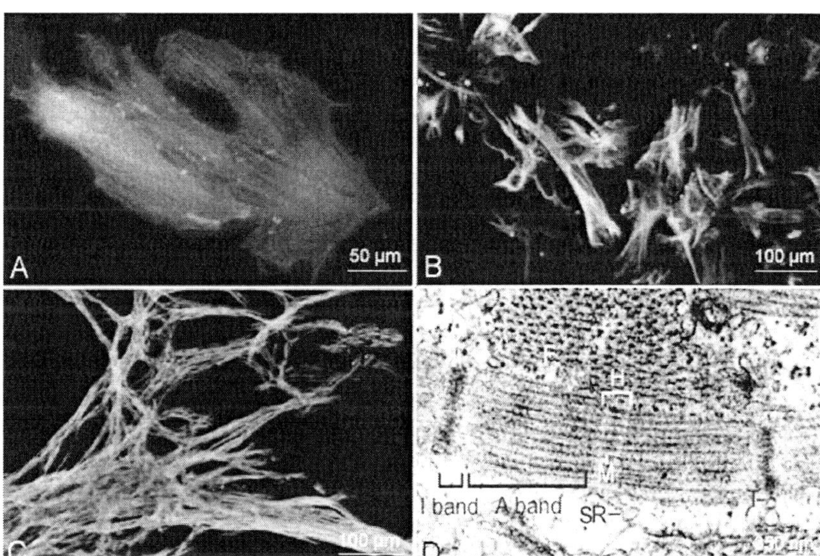

Figure 7. Expression of cardiomyocyte specific markers and ultrastructural analysis of ESC-derived cardiomyocytes. Immunocytochemistry was performed using (A) anti-α- actinin, (B) anti-desmin, (C) anti-troponin-I and (D) transmission electron microscopy of a sarcomere (adapted from Baharvand et al., 2005b), F: myofibrils, H: H-band, M: M-band, SR: sarcoplasmic reticulum, T: T-tubule, Z: Z-line.

The prevalence of cardiomyctes in whole cell populations of EBs is generally less than 10%. Thus improving directed differentiation events and improving the selection methods which enriches the population of cardiomyocytes has been achieved using genetic methodologies and cell sorting tools (Muller et al., 2000; Klug et al., 1996).

Cardiomyocyte differentiation can be enhanced in the mESC system following the addition of differentiation factors including, dimethyl sulfoxide (DMSO), RA (Wobus et al., 1997), and small molecules, such as Dynorphin B (Ventura et al., 2003) and cardiogenol derivatives (Wu et al., 2004). Addition of the demethylating agent 5-aza-2'-deoxycytidine to EB cultures has also been shown to be effective for mESC and hESC differentiation into cardiomyocytes. Interestingly, the effects of RA in hESCs do not appear to induce a higher proportion of cardiomyocytes in vitro (Schuldiner et al., 2000; Kehat et al., 2002). The addition of 5-aza-2'-deoxycytidine in later stages of EB growth is more effective in inducing cardiomycte differentiation compared with addition at earlier stages of EB culture (Xu et al., 2002a). Biological differences are thus apparent between murine and hESC (Kehat et al., 2002) differentiation into cardiomyctes and further studies are necessary to ascertain that hESC-derived cardiomyocytes may be suitable for cellular engrafting.

An alternative method for deriving cardiomyocytes has been achieved following the coculture of pluripotent hESC lines with END-2 cells (visceral-endoderm-like cell lines) Mummery et al. (2003). This was the first publication outlining the non-spontaneous formation of cardiomyocyte differentiation in hESCs and to compare the electrophysiological stage of hESC-derived cardiomyocytes with primary human fetal cardiomyocytes in vitro.

Cardiac transplantation studies have been conducted using ESC-derived cardiomyocytes from grafted EBS, although these contained a mixed population of different cardiac cell types (Klug et al., 1996; Min et al., 2002), although this would probably induce arrhythmogenicity and impair long-term cardiac function. Therefore, a prerequisite for using ESC-derived cardiomyocytes for the treatment of chronic heart failure or myocardial infarction would be to isolate only ventricular cardiomyocytes from EBs. One study has successfully shown that enrichment of hESC-derived cardiomyocytes can be achieved from differentiated cell cultures using a Percoll gradient density centrifugation strategy. This resulted in enriching the cell population for cardiomyocytes (70%) (Xu et al., 2002a). These cell populations proliferated in vitro and expressed specific cardiomyocyte markers.

Pure cardiomyocytic cultures have been achieved following the stable transfection of mESCs with a plasmid containing the neomycin resistance gene under control of the cardiac specific α-MHC-promoter (Hescheler et al., 1997). Selection using G418 resulted in a pure population of beating cardiomyocytes. Klug et al. (1996) transplanted these genetically manipulated mES-derived cardiomyocytes into the left ventricular wall of dystrophic mdx mice. Following cell transplantation, the donor cells were found to associate with the host cardiomyocytes. The G418-selected donor cardiomyocytes also displayed a typical differentiated cardiomyocytic phenotype, and were positive for myofibrillar proteins, MHC and the intermediate filament desmin. In ultrastructural analyses, the presence of intercalated discs was detected. This initial cell grafting study was repeated using a rat model displaying myocardial infarction (Min et al., 2002). Cardiomyocytes were obtained from the mESC line and transfected with the cDNA coding for green fluorescent protein (GFP). Following the intramyocardial engraftment, GFP-positive ES-derived cardiomyocytes were shown to survive in the infarcted rat myocardium and ameliorated heart function (Min et al., 2002). These donor cells appeared viable following the positive expression of anti-sarcomeric α-actin, which is not expressed in adult rat myocardium. Subsequent differentiation into mature cardiomyocytes could also be induced and was confirmed by positive expression of α-MHC and cardiac troponin I.

Other fluorescent reporters have also been used for targeting cardiomyocyte subtypes. Highly purified cardiomyocyte subtypes have been obtained for cell transplantation studies using other reporters. Muller et al. (2000) established a purification protocol based on the ventricular specificity of the myosin light chain-2 (MLC-2v) promoter fragment ventricular cardiomyocytes from cultivated mESCs. This was achieved using the enhanced green fluorescent protein (EGFP) tagged to the MLC-2v promoter. Electrophysiological analyses of beating fluorescent ESCs 21 days post differentiation indicated more than 80% of the contracting cardiomyocytes displayed ventricular properties. Enzymatic separation and application of a density gradient centrifugation step resulted in a three- to four fold enrichment of these EGFP-positive cardiomyocytes. Further cell sorting analyses resulted in a highly purified population consisting of 97–98% fluorescent cardiomyocytes (Muller et al., 2000). The purified cardiomyocytes contained the typical contractile fibers and an intact sarcomeric structure (Muller et al., 2000). It appears that EGFP may be a superior cassette compare with α-MHC- and the α-actin promoters, which are not specific for subtypes of cardiomyocytes (Klug et al., 1996; Kolossov et al., 1998). Isolation of cardiomyocytes using this EGFP targeting strategy and antibiotic selection may induce the expression of foreign proteins within the transfected cells which could induce immunological rejection if used for cell transplantation (Klug et al., 1996). In addition, the FACS analysis impairs the viability of ventricular cardiomyocytes as their electrophysiological properties and their contractile potential are lost, which may be attributed to cell stress. For this reason, alternative MLC-2v promoter-based approaches for the isolation of ventricular cardiomyocytes continues to be of major focus for stem cell cardiomyocyte research.

Differentiation of Embryonic Stem Cells into Islet-Like Structures

Insulin expression can be detected during early differentiation events in ESCs grown in suspension cultures, and which begin to form EBs. Insulin expression has been shown to be temporally and spatially expressed in the outer layer of EBs (corresponding to the embryonic endodermal layer) (Assady et al., 2001; Shiroi et al., 2002; Houard, et al., 2003; Moritoh et al., 2003). Insulin expression may also occur spontaneously during cultivation of undifferentiated ESCs (Soria et al., 2000). Clonal cells expressing insulin can be selected using cell-trapping methods and can be further differentiated into cell types that produce high levels of insulin (Soria et al., 2000). These clonal cells were induced to differentiate following culture in low levels of glucose and nicotinamide, which resulted in insulin production at a similar magnitude to β-cells and following cell transplantation, reversed the hyperglycemic phenotype of diabetic mice (Soria et al., 2000). This study was the first to demonstrate the potential of ESCs in curing diabetes.

In spite of this research, differentiation of ESCs into mature pancreatic β-cells in vitro has not been reproducible. In vitro directed differentiation of pancreatic islet-like structures from ESCs was first reported by Lumelsky et al. (2001). It has been reported that many similarities in development of the pancreas compared with development of the nervous system could be observed, even though these two organs develop from different germ layers (Kim et al., 1997; Li et al., 1999; Schwitzgebel et al., 2000). This notion was used to develop a strategy for generating neurons (Okabe et al., 1996) and to derive endocrine pancreatic cells from ESCs. The gene, nestin was recently identified in a subset of immature hormone-negative pancreatic cells (Zulewski et al., 2001) that, upon differentiation in vitro, give rise to insulin- and glucagon-expressing cells. The positive expression of endocrine pancreatic gene expression in the progeny of nestin- positive cell populations derived from ESCs was demonstrated. Final induction of differentiation into neurons and endocrine pancreatic cells was achieved following the addition of nicotinamide and removal of bFGF. Nicotinamide is a poly-synthetase (ADP-ribose) inhibitor known to induce cell differentiation and increase β-cell mass in cultured fetal pancreatic cells (Otonkoski, 1993). This technique was reproduced by Rajagopal et al. (2003), but for inducing differentiation of islet-like structures from mouse and hESCs. The studies of Sipione et al. (2004) and Hansson et al. (2004) complement this study with further characterization of the system and its potential for reversing the diabetic phenotype. In these studies, insulin-stained cells differentiated from ESCs were devoid of typical beta-cell granules and rarely showed immunoreactivity for C-peptide (a byproduct of de novo insulin synthesis cleaved from proinsulin during insulin processing) or were mostly apoptotic. The insulin release was variable and did not follow the anticipated release kinetics found in β-cells, suggesting that this hormonal secretion may not mimic the in vivo regulatory secretion response which is stimulated by glucose. Hori et al. (2002) selected and expanded nestin positive cells by using the protocol described by Lumelsky et al. (2001), however, in addition, the cells were treated with LY294002 (an inhibitor of phosphoinositide 3-kinase (PI3K)) and an essential intracellular signaling regulator, which resulted in the production of cells that resembled pancreatic β-cells. The differentiated ESCs in this system produced insulin at levels far greater than previously reported and displayed the ability to release insulin via the glucose response. Cells cultured in the presence of a phosphoinositide 3-kinase inhibitor, were not C-peptide immunoreactive, and absorbed fluorescein isothiocyanate (FITC)-labeled insulin from the culture medium (Hansson et al., 2004).

Molecular expression profiling has helped to further improve the differentiation of pancreatic cells such as β-cells from ESCs in vitro. Blyszczuk et al. (2003) showed that constitutive expression of Pax4 affects the differentiation of ESCs and significantly promotes the development of insulin producing cells. In Pax4-overexpressing ESCs, isl-1, ngn3, insulin, islet amyloid polypeptide, and glucose transporter 2 (Glut-2) mRNA levels are upregulated (Blyszczuk et al., 2003). Transplantation of wild type (wt) and $Pax4^+$ insulin-producing cells was sufficient for preserving the normal range of blood glucose levels in diabetic mice. One plausible explanation for this could be due to the different locations of the implanted cells i.e., the kidney capsule and spleen (Blyszczuk et al., 2003) versus subcutaneous injection (Lumelsky et al., 2001). However, only a marginal effect of pdx-1 expression was observed following cell transplantation. It has also been proposed that this may be due to differences in the expression levels of pdx-1 within pancreatic cells (Miyazaki et al., 2004). Research has indicated that high-levels of pdx-1 expression in undifferentiated ESCs is toxic and results in cell death (Miyazaki et al., 2004).). A more recent report avoided the use of ITSFn and FGF-2 in culture medium, which does not select or enrich for specific cell types prior to pancreatic cell differentiation (Blyszczuk et al., 2004). Functional beta-like cells were produced by differentiating EBs into multi-lineage progenitor cells and subsequently inducing pancreatic differentiation using serum-free medium containing nicotinamide and laminin. Islet-like clusters were characterized by expression of C-peptide and insulin. Cells isolated from islet-like clusters show glucose-dependent insulin release at terminal stages of development. Blyszczuk et al., was the first to demonstrate that ES-derived pancreatic cells had some functional properties including ionic channels which are observed in embryonic beta-like cells.

A novel method for inducing ESCs to differentiate into insulin-producing cells was introduced by using the *Nkx6.1* promoter (Leon-Quinto et al., 2004). This was achieved following the elimination of LIF from the culture medium, combined with reducing serum concentrations (from 15% to 3%) and with the addition of selected factors. These factors included anti-sonic hedgehog and nicotinamide. Cells were then co-cultured with embryonic pancreases, to induce the formation of islets. Islets expressing Nkx6.1 were selected by using FACS or cell trapping. Approximately 20% of the cell cultures co-expressed insulin and Pdx-1. Cell trapping selection yielded an almost pure population of insulin-positive cells, which expressed the beta cell genes/proteins Pdx-1, Nkx6.1, insulin, glucokinase, GLUT-2 and Sur-1. Transplantation of these cells into streptozotocin-induced diabetic mice normalized their glycaemia over a period of days.

Recently, a modification of Lumelskey and colleagues (2001) method resulted in the production of insulin-secreting cells derived from hESCs (Segev et al., 2004). This was achieved following an additional step of culture including, a lowering of the glucose concentration in the medium, removal of bFGF and addition of nicotinamide. Dissociating the cells and growing them in suspension resulted in the formation of clusters which secreted higher levels of insulin than their in vivo counterparts and could be maintained in vitro. These cells expressed pancreatic genes and following immunofluorescence and in situ hybridization studies, it was confirmed that a high percentage of insulin-expressing cells were located within these cell clusters.

A recent report has shown that ES-derived embryoid bodies cultured in the presence of stage-specific concentrations of monothioglycerol (MTG) and 15% fetal calf serum, followed by serum-free conditions, gave rise to a population of cells that positively express insulin I, insulin II, pdx-1, and Sox17 (an endoderm marker) (Ku et al., 2004). Further immunohistochemical analysis of these cells indicated the presence of intracellular insulin particles, and its de novo production was confirmed by staining for C-peptide. The insulin positive or C-peptide positive cells also coexpressed glucagon, thus demonstrating these cells may be pancreatic precursors rather than precursors of yolk sac or fetal liver. The addition of β-cell specification and differentiation factors, activin-B, nicotinamide, and beta cell growth factor, exendin-4 (Xu et al., 1999; Tourrel et al., 2001; Movassat et al., 2002; Tourrel et al., 2002) to later-stage cultures increased the number of insulin-positive cells to 2.73% of the total population, compared with the control culture, which gave rise to less than 1% insulin-staining cells. In contrast, Shi et al. (2005) demonstrated that a simple three-step experimental approach, based on the combination of activin A, all trans-retinoic acid, and other mature factors could induce mESCs to differentiate into insulin-producing cells within 2 weeks, and that insulin release of these cells was regulated by altered levels of glucose. This insulin-enhanced green fluorescent green protein reporter system excludes the possibility of insulin uptake. Transplantation of these ESC derived insulin-positive cells could normalize blood glucose levels and rescue the survival of streptozocin-induced diabetic mice.

Lester et al. (2004) have recently differentiated non-human primate (rhesus) ESC lines into insulin producing, beta-like cells with exendin. To facilitate the identification of beta like cells and to track the cells post- transplantation, they developed a marker gene construct by fusing the human insulin promoter (HIP) to GFP (HIP-GFP). Fluorescent markers such as GFP are useful for the purification of specific cell populations by using FACS as well as to track live expression subsequent to cell transplantation. This construct was transfected into rhesus ESC-derived cells and GFP expression was identified (representative of C-peptide positive cells), thereby substantiating endogenous insulin production by rhesus ESC-derived cells. This approach facilitates the purification of insulin producing cells from mixed cell populations and to track these cells post-transplantation in the non-human primate model (Eiges et al., 2001).

Embryonic Stem Cells for Regenerative Medicine

Stem cells have the potential to be used for a number of different purposes: (1) investigation of the mechanisms involved in early mammalian specific tissue lineage commitment, differentiation, and maturation; (2) derivation of a unique in vitro model to study mammalian specific tissue; (3) functional genomics, drug and growth factor discovery, drug testing, and reproductive toxicology; and (4) development of tissue-engineering strategies, but the most promising is (5) restore or replace diseased tissue such as myocardial infarction, Parkinson's disease, diabetes, spinal cord injury, aging and etc.

Replacement of worn out or injured cells with functional, healthy cells may restore the normal function of the tissues or organs. This underlying principle represents the holistic approach of cell replacement therapy, or regenerative medicine. It is hoped that organs or tissues treated may function normally and more efficiently than tissues treated using conventional therapies such as injection of recombinant insulin to treat diabetes or replacement of organs using mechanical devices such as pacemakers in the heart. Embryonic SCs are anticipated to become a powerful tool for future regenerative medicine due to their capacity of self-renewal and pluripotency. Studies in animal models have shown that transplantation of fetal stem cell, ESCs, or pluripotent stem cell derivatives can successfully treat many chronic diseases, such as Parkinson disease, diabetes, spinal cord injury, Purkinje cell degeneration, Duchenne muscular dystrophy, liver or heart failure, and osteogenesis imperfecta (Zhang et al., 1996; McDonald et al., 1999; Horwitz et al., 1999; Soria et al., 2000; Kobayashi et al., 2000; Li et al., 2000; Kim et al., 2002).

To achieve an effective, long-lasting, and stable repair of damaged or diseased tissues with hESCs, a number of 'basic' criteria must be satisfied. These include: (1) Existing hESC lines are not suitable for human use as the majority of these lines were derived on animal feeder layers. However, the feeder layer is required to support the proliferation and prevent the differentiation of pluripotent stem cells. Direct contact with mouse feeders renders these lines xenotransplantation products. In addition, materials derived from animal species and reagents that are not compliant with the requirements of regulatory authorities were used to develop and propagate the existing hESC lines. New cell lines derived under stringent regulatory conditions, in an animal-free culture system with human feeders or in feeder-free conditions, are required. The culture systems that are currently used to propagate hESCs are far from ideal. The propagation of stem cells is feeder dependent, or requires medium conditioned by feeders. Recently, Ludwig et al., (2005) reported feeder-independent hESC culture that includes protein components solely derived from recombinant sources or purified from human material. They described the derivation of two new human ES cell lines in these defined culture conditions. (2) Identifying a common set of molecular markers and understanding how hESC lines from diverse genetic backgrounds and which are derived in different labs differ from each other. Indeed, there appear to be several significant differences at the molecular level between different hESC lines (Richards et al., 2004). Some hESC lines seem predisposed to differentiate along a particular cell lineage and spontaneously differentiate, for example, into cardiomyocytes, whereas other hESC lines might more readily form neural precursors (Abeyta et al., 2004). (3) The karyotypic stability of hESC lines over hundreds of population doublings in vitro, and the frequency of occurrence of aneuploidy, need to be accurately assessed. (4) Generating large-scale production of cells for cell transplantation and tissue repair. Standardized bulk cultures and maintaining the proliferation of pure undifferentiated hESCs are essential for developing specialized cells for transplantation. (5) Differentiating cells towards the correct phenotype and which are functional. (6) Ensuring cells/tissues adopt the appropriate three dimensional organization and produce ECM (with provision of structural support if required: resorbable scaffold). (7) Producing cells/tissues that are structurally and mechanically compliant with the normal demands of native tissue. (8) Achieving full integration with native tissue such as vascularization (if required) and overcoming the risk of immunological rejection. (9) Ethical, religious, and political issues associated with derivation and use of hESCs have not been fully resolved and remain stumbling blocks that impede scientific investigation in this area. However, to get around this, Chung et al., (2006) have adapted a method commonly used in assisted-reproduction clinics for preimplantation genetic diagnosis in mouse. This involves removing a cell from the eight-cell stage of development this 'blastomere' cell is then analysed for genetic defects. Instead, they used the blastomere cell to produce ESC lines — without compromising the embryo from which the blastomere was obtained. The single blastomeres are co-cultured with established ES cell lines, and then separated from them to form fully competent ESC lines.

Therefore, before any clinical application of ESCs can succeed, a number of obstacles must first be overcome. Since hESCs spontaneously form teratomas in vivo, the first obstacle is their tumorigenic potential. Therefore, establishing a highly efficient purification protocol is crucial for generating a specialized cell type. Negative selection of Oct-4 expressing cells could be employed to tackle this issue. New strategies and methodologies are needed to isolate terminally differentiated cells. Implants of ESCs could be tagged with apoptotic signals which are triggered if unspecialized differentiation occurs, or if severe complications arise. New methodologies and inevitable new discoveries will overcome these apparent hurdles. To date, no strategies employed in ESC differentiation has yielded a 100% pure population of mature progeny. This is essential otherwise transplantation of undifferentiated ESCs or inappropriate cell lineages may heighten the risk of teratoma formation or further perturbation of tissue function. Therefore, there efficient protocols for purifying cell populations are required. Methods such as fluorescence-activated cell sorting (FACS) or magnetic-activated cell sorting (MACS) allow such purification but are dependent on the cell type of interest expressing a surface marker that can be recognized by a fluorescent or magnetic microbead-tagged antibody and, to be fully effective, the marker needs to be cell-type specific. In most cases, these cell markers are not commercially available, thus, sorting methods are reliant on genetic modification of the ESCs by tagging a lineage-specific promoter to a fluorescent marker. Some protocols employ selective culture conditions which promote the growth of one particular cell type at the expense of the other. Alternatively, cells could be transduced with a drug-resistance gene instead of a marker, to allow for preferential selection of subpopulations. The drawback of these two approaches is that the genetically manipulated ESC derivatives might be immunologically rejected since they express the foreign proteins as well as the selection and fluorescent markers.

Also of concern is the potential of hESCs to transmit infections, but this only a marginal concern, if adequate derivation and culture standards can be achieved. Another major objective will be to develop ways to overcome immunological rejection of ESCs. Several approaches have been proposed which may help overcome this problem: (1) Minimizing the donor-recipient alloantigenic differences by either establishing a stem-cell bank. In this case and given the large number of supernumerary IVF embryos available world wide, deriving and maintaining banks of HLA-typed hESC lines from different genetic and ethnic backgrounds might be a more feasible solution in overcoming the tissue rejection problem. (2) Nuclear transfer or therapeutic cloning or genomic replacement is one approach in obtaining rejection-free, customized 'tailor made' hESC lines. The nucleus from a normal somatic cell obtained from the recipient, such as from a skin biopsy, is injected into an enucleated oocyte. The cytoplasm of this oocyte has the ability to reprogram the differentiated nucleus and to re-establish the embryonic genome. The blastocyst formed from this oocyte would be the source for the derivation of new ESC lines which would be genetically matched for each nuclear gene of the recipient (Hochedlinger et al., 2003; Hochedlinger et al., 2004). The derivation of a human blastocyst after heterologous nuclear transfer to donated oocytes was achieved recently by Stojkovic et al., (2005a). To overcome political, ethical, and religious grounds, a possible solution to the controversy, proposed by many who want the medical science to progress, might be to invent a process that produces an entity that cannot implant in the uterus-termed alternative nuclear transfer (ANT) by William Hurlbut (2004), a member of President Bush's Council on Bioethics. Meissner and Jaenisch (2006) have now developed a method to accomplish ANT in mouse. Their technique builds on previous work by Strumpf et al., (2005) who studied a gene called *cdx2* and its role in establishing the mouse trophectoderm and, later, the intestinal tract. Their results suggested that if this gene was suppressed in the nucleus of the donor cell during the NT process, it might allow the generation of NT entities that could not implant. Meissner and Jaenisch (2006) demonstrated that this was indeed the case, using a clever method to control *cdx2* expression at various stages. They introduced into the donor cell a gene encoding an RNA that inhibits *cdx2* expression, and this gene was transmitted with the donor nucleus to the egg and continued to be active during the NT. Once they had derived the ANT pluripotent stem-cell line from the resulting embryoid blastocysts, they clipped out the inhibitor gene to enable the resulting ANT stem cells to produce mature intestinal epithelia given the right cues. These ANT pluripotent stem-cell lines can form many other mature cells, just as the classical ES and NT cell lines do.

At present, however, the widespread use of nuclear transfer technology to create such cell lines for individual patients seems untenable because of the extremely low success rates of a highly inefficient procedure, the paucity of donor human oocyte and the unknown repercussions of absence of the sperm-imprinting mechanisms (Mann, 2001). In this case, whole MHC regions in ESCs and other immune molecules will be identical to that of the patient, except for the mitochondrial genome. Using therapeutic cloning may therefore by most efficient for overcoming the hurdle of immune rejection for hESC replacement therapies. Although, several ethical problems should be solved by social consensus before clinical studies in this direction can be initiated. (3) Immunosuppressive therapy; (4) Genetic manipulation (to create a 'universal-donor'). This could be achieved by insertion of immunosuppressive molecules such as Fas ligand, or by deleting immunoreactive molecules such B7 antigens (Walker et al., 1997; Harlan and Kirk, 1999). More ambitiously, foreign major histocompatability complex (MHC) genes could be replaced with host MHC genes so that the host immune system will recognize the implanted cells as 'self' (Westphal and Leder, 1997; Hardy and Malissen, 1998), thus increasing the immunocompatability of the cells. This approach is could be almost feasible if combined with homologous recombination. Non-immune adult somatic cells express only MHC class I antigens, whereas cells of the immune system also express MHC class II antigens. Very low expression levels of MHC class I proteins were recently demonstrated on the surface of hESCs. The expression was moderately increased following in vitro or in vivo differentiation. The MHC class II antigens were not expressed by undifferentiated or differentiated cells (Drukker et al., 2002). Therefore, in contrast to transplanted solid organs, which harbor MHC class II expressing antigen presenting cells (e.g. macrophages, dendritic cells, lymphocytes), homogenous non-immune cell populations differentiated from ESCs, are expected not to express MHC class II antigens and might provoke a lesser degree of host rejection (Odorico et al., 2001). Nevertheless, even theoretically, the effect of MHC expression alterations is limited. None of these genetic manipulations are expected to decrease non-MHC-mediated rejection. (5) Induction of donor specific tolerance (haematopoietic chimerism; donor-blood pre-transplantation; lymphocyte infusion pre-transplant). By using the same ESC lines to derive both hematopoietic stem cells and other lineages, it may be possible to initially achieve hematopoietic chimerism followed by engraftment of a second cell type. Therefore, a second lineage should not be rejected as it would be regarded as "self" by the chimeric patient's bone marrow and immune system, which were derived, in part, from the same ESC line. This approach may be useful, as rat embryonic-like stem cells when injected into full MHC-mismatched rats, result in permanent engraftment that enabled further long-term graft acceptance of second-set transplantation via mixed chimerism (Fandrich et al., 2002).

Summary

The potential of ESCs to generate any cell, tissue or organ 'on demand' for tissue repair or for replacement promises to revolutionize the field of developmental biology, drug discovery and for the treatment of human disease. However, it should be realized that we are a long way off from the application of specialized cells for cell therapeutics and for the treatment of disease. This will occur if the hurdles associated with current ESC growth and differentiation is overcome.

Acknowledgment

The author wish to sincerely thank colleagues at department of stem cells in Royan institute including Dr M. Ebrahimi, A. Farrokhi, Dr H. Gourabi, M. Hashemi, M. Hatami, H. Jafary, L. Khodadadi, Miss Kiani, M. Massumi, S. Mollamohammadi, S. Z. Moradi, Miss Namiri, Dr M. Rezazadeh Valojerdi, M. Pakzad, M. Pirouz, Dr A. Shahverdi, A. Taee, Mr Taghiabadi, Dr A. Vosough, N. Zare and Dr Vanessa Hall (University of Newcastle, UK) for critically reading the manuscript.

References

Abbondanzo, SJ; Gadi, I; Stewart, CL. Derivation of embryonic stem cell lines. In: Wasserman, PM; Depamphilis, ML, ed. *Methods enzymology*. San Diego CA: Academic Press; 1993; 803–823.

Abeyta, MJ; Clark, AT; Rodriguez, RT; Bodnar, MS; Pera, RA; Firpo, MT.Unique gene expression signatures of independently-derived human embryonic stem cell lines. *Hum Mol Genet*, 2004 13,601–608.

Adamson, ED; Minchiotti, G; Salomon, DS. Cripto: a tumor growth factor and more. *J Cell Physiol,* 2002 190, 267-278.

Amit, M; Carpenter, MK; Inokuma, MS; Chiu,CP;Harris,CP; Waknitz, MA; Itskovitz-Eldor,J; Thomson,JA. Clonally derived human embryonic stem cell lines maintain pluripotency and proliferative potential for prolonged periods of culture. *Dev Biol*, 2000 227,271–278.

Amit, M; Margulets, V; Segev,H; Shariki,K; Laevsky, I; Coleman,R; Itskovitz-Eldor, J. Human feeder layers for human embryonic stem cells. *Biol Reprod*, 2003 68, 2150-6.

Amit, M; Shariki,C; Margulets,V; Itskovitz-Eldor, J. Feeder layer- and serum-free culture of human embryonic stem cells. *Biol Reprod*, 2004 70, 837-45.

Andrews, PW; Bronson, DL; Benham, F; Strickland, S; Knowles, BB. A comparative study of eight cell lines derived from human testicular teratocarcinoma. *Int J Cancer*, 1980 26, 269–280.

Andrews, PW; Damjanov, I; Simon, D; Banting, GS; Carlin, C; Dracopoli, NC; Fogh, J. Pluripotent embryonal carcinoma clones derived from the human teratocarcinoma cell line Tera-2: differentiation in vivo and in vitro. *Lab Invest*, 1984 50, 147–162.

Anisimov, SV; Tarasov, KV; Tweedie, D; Stern, MD;Wobus, AM; Boheler, KR. SAGE identification of gene transcripts with profiles unique to pluripotent mouse R1 embryonic stem cells. *Genomics*, 2002 79, 169–176.

Aprelikova, O; Pace, AJ; Fang, B; Koller, BH; Liu, ET. BRCA1 is a selective co-activator of 14-3-3 sigma gene transcription in mouse embryonic stem cells. *J Biol Chem*, 2001 276, 25647–25650.

Assady, S; Maor, G; Amit, M; Itskovitz-Eldor, J; Skorecki, KL; Tzukerman, M. Insulin production by human embryonic stem cells. *Diabetes*, 2001 50, 1691-7.

Auerbach,W; Dunmore, JH; Fairchild-Huntress, V; Fang, Q; Auerbach, AB; Huszar, D; Joyner, AL. Establishment and chimera analysis of 129/SvEv- and C57BL/6-derived mouse embryonic stem cell lines. *BioTechniques*, 2000 29, 1024–1032.

Avilion, AA; Nicolis, SK; Pevny, LH; Perez, L; Vivian, N; Lovell-Badge, R. Multipotent cell lineages in early mouse development depend on SOX2 function. *Genes Dev*, 2003 17, 126–140.

Baharvand, H; Azarnia, M; Parivar, K; Kazemi Ashtiani, S. The effect of extra cellular matrix on embryonic stem cell-derived cardiomyocytes. *J Mol Cell Cardiol*, 2005 38, 495-503.

Baharvand, H; Kazemi Ashtiani, S; Taee, A; Massumi, M; Valojerdi, MR; Eftekhari, P; Zarei Zarei Moradi, Sh. Geration of New Human Embryonic Stem Cell Lines with Diploid and Triploid Karyotypes. *Development, Growth and Differentiation*, 2006 48, 117-28.

Baharvand, H; Kazemi Ashtiani, S; Valojerdi, MR; Shahverdi, A; Taee, A; Sabour, D. Establishment and in vitro differentiation of a new embryonic stem cell line from human blastocyst. *Differentiation*, 2004a 72, 224–9.

Baharvand, H; Matthaei, KI. Culture condition difference for establishment of new embryonic stem cell lines from the C57BL/6 and BALB/c mouse strains. *In Vitro Cell Dev Biol Anim,* 2004 40, 76-81.

Baharvand, H; Power, JM; Ozsarac, N; Matthaei, KI. Differentiation of embryonic stem cells into functional neurons in vitro. *Neuroscience Research Communications*, 2004b 35, 130-8.

Bain, G; Kitchens, D; Yao, M; Huettner, JE; Gottlieb, DI. Embryonic stem cells express neuronal properties in vitro. *Dev Biol*, 1995 168, 342-57.

Barberi, T; Klivenyi, P; Calingasan, NY; Lee, H; Kawamata, H; Loonam, K; Perrier, AL; Bruses, J; Rubio, ME;Topf,N; Tabar, V; Harrison, NL; Beal, MF; Moore, MA; Studer, L. Neural subtype specification of fertilization and nuclear transfer embryonic stem cells and application in parkinsonian mice. *Nat Biotechnol*, 2003 21, 1200-7.

Beattie, GM; Lopez, AD; Bucay, N; Hinton, A; Firpo, MT; King, CC; Hayek, A. Activin A maintains pluripotency of human embryonic stem cells in the absence of feeder layers. *Stem Cells*, 2005 23, 489-95.

Beck, KD; Valverde, J; Alexi, T; Poulsen, K; Moffat, B; Vandlen, RA; Rosenthal, A; Hefti, F. Mesencephalic dopaminergic neurons protected by GDNF from axotomy-induced degeneration in the adult brain. *Nature*, 1995 373, 339-41.

Ben-Shushan, E; Thompson, JR; Gudas, LJ; Bergman, Y. Rex-1, a gene encoding a transcription factor expressed in the early embryo, is regulated via Oct-3/4 and Oct-6 binding to an octamer site and a novel protein, Rox-1, binding to an adjacent site. *Mol Cell Biol,* 1998 18, 1866–1878.

Bhattacharya, B; Miura, T; Brandenberg, R; Mejido, J; Luo, Y; Yang, AX; Joshi, BH; Irene, G; Thies, RS; Amit, M; Lyons, I; Condie, BG; Iskovitz-Eldor, J; Rao, MS; Puri, RK. Gene expression in human embryonic stem cell lines: unique molecular signature. *Blood*, 2004 103, 2956–2964.

Bibel, M; Richter, J; Schrenk, K; Tucker, KL; Staiger, V; Korte, M; Goetz, M; Barde, YA. Differentiation of mouse embryonic stem cells into a defined neuronal lineage. *Nat Neurosci,* 2004 7, 1003-9.

Blyszczuk, P; Asbrand, C; Rozzo, A; Kania, G; St-Onge, L; Rupnik, M; Wobus, AM. Embryonic stem cells differentiate into insulin-producing cells without selection of nestin-expressing cells. *Int J Dev Biol,* 2004 48,1095-104.

Blyszczuk, P; Czyz, J; Kania, G; Wagner, M; Roll, U; St-Onge, L; Wobus, AM. Expression of Pax4 in embryonic stem cells promotes differentiation of nestin-positive progenitor and insulin-producing cells. *Proc Natl Acad Sci USA*, 2003 100, 998–1003.

Bongso, A; Fong, CY; Ng, SC; Ratnam, S. Isolation and culture of inner cell mass cells from human blastocysts. *Hum Reprod*, 1994 9, 2110-7.

Bradley, A; Evans, MJ; Kaufman, MH; Robertson, E.Formation of germline chimaeras from embryo-derived teratocarcinoma cell lines. *Nature*, 1984 309, 255–256.

Branton, RL; Love, RM; Clarke, DJ. cAMP included during cell suspension preparation improves survival of dopaminergic neurons in vitro. *Neuroreport*, 1998 9, 3223-7.

Brenin, D; Look, J; Bader, M; Hubner, N; Levan,G; Iannaccone, P. Rat embryonic stem cells: a progress report. *Transplant Proc*, 1997 29, 1761–1765.

Brimble, SN; Zeng, X; Weiler, DA; Luo,Y; Liu, Y; Lyons, IG; Freed, WJ; Robins, AJ; Rao, MS; Schulz, TC. Karyotypic stability, genotyping, differentiation, feeder-free maintenance, and gene expression sampling in three human embryonic stem cell lines derived prior to August 9, 2001. *Stem Cells Dev*. 2004 13, 585-97.

Brinster, RL. The effect of cells transferred into the mouse blastocyst on subsequent development. *J Exp Med*, 1974 140, 1049–56.

Brook, FA; Gardner, RL. The origin and efficient derivation of embryonic stem cells in the mouse. *Proc Natl Acad Sci USA*, 1997 94, 5709–5712.

Buehr, M; Nichols, J; Stenhouse, F; Mountford, P; Greenhalgh, CJ; Kantachuvesiri, S; Brooker, G; Mullins, J; Smith, AG. Rapid loss of Oct-4 and pluripotency in cultured rodent blastocysts and derivative cell lines. *Biol Reprod*, 2003 68, 222–229.

Burdon, T; Smith, A; Savatier, P. Signalling, cell cycle and pluripotency in embryonic stem cells. *Trends Cell Biol,* 2002 12, 432–438.

Burdon, T; Stracey, C; Chambers, I; Nichols, J; Smith, A. Suppression of SHP-2 and ERK signalling promotes self-renewal of mouse embryonic stem cells. *Dev Biol*, 1999 210, 30–43.

Buzzard, J J; Gough, NM; Crook, JM; Colman, A. Karyotype of human ES cells during extended culture. *Nat Biotechnol*, 2004 22, 381-2.

Cai, J; Cheng, A; Luo,Y; Lu, C; Mattson, MP; Rao, MS; Furukawa, K. Membrane properties of rat embryonic multipotent neural stem cells.*J Neurochem*, 2004 88,212– 226.

Carpenter, MK; Rosler, E; Rao, MS. Characterization and differentiation of human embryonic stem cells. *Cloning Stem Cells*,2003 5, 79– 88.

Carpenter, MK; Rosler, ES; Fisk, GJ; Brandenberger, R; Ares, X; Miura, T; Lucero M, Rao MS. Properties of four human embryonic stem cell lines maintained in a feeder-free culture system. *Dev Dyn*, 2004 229, 243-58.

Caviedes, P; Olivares, E; Salas, K; Caviedes, R; Jaimovich, E. Calcium fluxes, ion currents and dihydropyridine receptors in a new immortal cell line from rat heart muscle. *J Mol Cell Cardiol,* 1993 25, 829-845.

Chacko, KJ.Observations on the ultra structure of developing myocardium of rat embryos. *J Morphol*, 1976 150, 681–709.

Chambers, I; Colby, D; Robertson, M; Nichols, J; Lee, S; Tweedie, S; Smith, A. Functional expression cloning of nanog, a pluripotency sustaining factor in embryonic stem cells. *Cell*, 2003 113, 643–655.

Chapman, G; Remiszewski, JL; Webb, GC; Schulz, TC; Bottema, CD; Rathjen, PD. The mouse homeobox gene, *Gbx2*: genomic organization and expression in pluripotent cells in vitro and in vivo. *Genomics,* 1997 46, 223–233.

Chen, Y; He, ZX; Liu, A; Wang, K; Mao, WW; Chu, JX; Lu, Y; Fang, ZF; Shi, YT; Yang, QZ; Chen da, Y; Wang, MK; Li, JS; Huang, SL; Kong, XY; Shi, YZ; Wang, ZQ; Xia, JH; Long, ZG; Xue, ZG; Ding, WX; Sheng, HZ. Embryonic stem cells generated by nuclear transfer of human somatic nuclei into rabbit oocytes. *Cell Res*, 2003 13, 251-63.

Chen, H; Qian, K; Hu, J; Liu, D; Lu, W; Yang, Y; Wang, D; Yan, H; Zhang, S; Zhu, G. The derivation of two additional human embryonic stem cell lines from day 3 embryos with low morphological scores. *Hum Reprod.* 2005 20, 2201-2206.

Cheng, L; Hammond, H; Ye, Z; Zhan, X; Dravid, G. Human adult marrow cells support prolonged expansion of human embryonic stem cells in culture. *Stem Cells*, 2003 21, 131– 42.

Choo, AB; Padmanabhan, J; Chin, AC; Oh, SK. Expansion of pluripotent human embryonic stem cells on human feeders. *Biotechnol Bioeng*, 2004 88, 321-31.

Chung Y, Klimanskaya I, Becker S, Marh J, Lu SJ, Johnson J, Meisner L, Lanza R. Embryonic and extraembryonic stem cell lines derived from single mouse blastomeres. Nature. 2006 439:216-219.

Cibelli, JB; Stice, SL; Golueke, PG; Kane, JJ; Jerry, J; Blackwell, ESC; Ponce de Leon, FA; Robl, JM. Transgenic bovine chimeric offspring produced from somatic cell-derived stem-like cells. *Nat Biotechnol*, 1998 16, 642-646.

Clark, AT; Bodnar, MS; Fox, M; Rodriquez; RT, Abeyta; MJ, Firpo; MT, Pera; RA. Spontaneous differentiation of germ cells from human embryonic stem cells in vitro. *Hum Mol Genet*, 2004 13, 727-39.

Cowan, CA; Klimanskaya, I; McMahon, J; Atienza, J; Witmyer, J; Zucker, JP; Wang, S; Morton, CC; McMahon, AP; Powers, D; Melton, DA. Derivation of embryonic stem-cell lines from human blastocysts. *N Engl J Med.* 2004 350, 1353-6.

D'Amour, KA; Gage, FH. Genetic and functional differences between multipotent neural and pluripotent embryonic stem cells. *Proc Natl Acad Sci USA*, 2003 100, 11866–11872.

Daheron, L; Opitz, SL; Zaehres, H; Lensch, WM; Andrews, PW; Itskovitz-Eldor, J; Daley, GQ. LIF/STAT3 signaling fails to maintain self-renewal of human embryonic stem cells. *Stem Cells*, 2004 22, 770-8.

Dani, C; Chambers, I; Johnstone, S; Robertson, M; Ebrahimi, B; Saito, M; Taga,T; Li, M; Burdon, T; Nichols, J; Smith, A. Paracrine induction of stem cell renewal by LIF-deficient cells: a new ES cell regulatory pathway. *Dev Biol,* 1998 203, 149-162.

De Haan, G. Hematopoietic stem cells: self-renewing or aging? *Cells Tissues Organs*, 2002 171, 27–37.

Delhasie, F; Bralion, V; Schuurbiers, N; Dessy, F. Establishment of an embryonic stem cell line from 8-cell stage mouse embryos. *Eur. J. Morphol.* 1996 34, 237-243.

Diwan, SB; Stevens, LC. Development of teratomas from ectoderm of mouse egg cylinders. *J Natl Cancer Inst*, 1976 57, 937–42.

Doetschman, T; Williams, CP; Maeda, N. Establishment of hamster blastocyst - derived embryonic stem (ES) cells. *Dev Biol*, 1988 127, 224-227.

Downing, GJ; Battey, JF Jr. Technical assessment of the first 20 years of research using mouse embryonic stem cell lines. *Stem Cells,* 2004 22, 1168-80.

Draper, JS; Smith, K; Gokhale, P; Moore, HD; Maltby, E; Johnson, J; Meisner, L; Zwaka TP, Thomson JA, Andrews PW. Recurrent gain of chromosomes 17q and 12 in cultured human embryonic stem cells. *Nat Biotechnol.* 2004 22, 53-4.

Drukker, M; Katz, G; Urbach, A; Schuldiner, M; Markel; G; Itskovitz-Eldor, J; Reubinoff, B; Mandelboim, O; Benvenisty, N.Characterization of the expression of MHC proteins in human embryonic stem cells. *Proc Natl Acad Sci U S A*. 2002 99, 9864-9

Easterday, MC; Dougherty, JD; Jackson, RL; Ou, J; Nakano, I; Paucar, AA; Roobini, B; Dianati, M; Irvin, DK; Weissman, IL; Terskikh, AV; Geschwind, DH; Kornblum, HI. Neural progenitor genes. Germinal zone expression and analysis of genetic overlap in stem cell populations. *Dev Biol*, 2003 264, 309–322.

Eggert, K; Schlegel, J; Oertel, W; Wurz, C; Krieg, JC; Vedder, H. Glial cell line-derived neurotrophic factor protects dopaminergic neurons from 6-hydroxydopamine toxicity in vitro. *Neurosci Lett*, 1999 269,178-82.

Eiges, R; Schuldiner, M; Drukker, M; Yanuka, O; Itskovitz-Eldor, J; Benveniste, H. Establishment of human embryonic stem celltransfected clones carrying a marker for undifferentiated cells. *Current Biology*, 2001 11, 514-8.

Eistetter, HR. Pluripotent embryonic stem cell lines can be established from disaggregated mouse morulae. *Dev Growth Differ*, 1989 31, 275–282.

Evans, MJ; Kaufman, MH. Establishment in culture of pluripotential cells from mouse embryos. *Nature*, 1981 292, 154–156.

Fan, Y; Melhem, MF; Chaillet, JR. Forced expression of the homeobox-containing gene Pem blocks differentiation of embryonic stem cells. *Dev Biol*, 1999 210, 481–496.

Fandrich, F; Lin, X; Chai, GX; Schulze, M; Ganten, D; Bader, M; Holle, J; Huang, DS; Parwaresch, R; Zavazava, N; Binas, B. Preimplantation-stage stem cells induce long-term allogeneic graft acceptance without supplementary host conditioning. *Nature Medicine*, 2002 8, 171–178.

Field, LJ. Transgenic mice in cardiovascular research. *Annu Rev Physiol*, 1993 55, 97-114.

Findikli, N; Kahraman, S; Akcin, O; Sertyel, S; Candan, Z. Establishment and characterization of new human embryonic stem cell lines. *Reprod Biomed Online.* 2005 10, 617-27.

Fortunel, NO; Otu, HH; Ng, HH; Chen, J; Mu, X; Chevassut, T; Li, X; Joseph, M; Bailey, C; Hatzfeld, JA; Hatzfeld, A; Usta, F; Vega, VB; Long, PM; Libermann, TA; Lim, B. Comment on 'Stemness': transcriptional profiling of embryonic and adult stem cells'' and ''a stem cell molecular signature.'' *Science*, 2003 302,393.

Fraichard, A; Chassande, O; Bilbaut, G; Dehay, C; Savatier, P; Samarut, J. In vitro differentiation of embryonic stem cells into glial cells and functional neurons. *J Cell Sci*, 1995 108, 3181-8.

Furuya, M; Yasuchika, K; Mizutani, K; Yoshimura, Y; Nakatsuji, N; Suemori, H. Electroporation of cynomolgus monkey embryonic stem cells.*Genesis,*2003 37, 180–187.

Gardner, DK; Lane, M. Towards a single embryo transfer. *Reproductive BioMedicine Online*, 2003 6, 470–481.

Gardner, DK; Surrey, E; Minjarez, D; Leitz, A; Stevens, J; Schoolcraft, WB. Single blastocyst transfer: a prospective randomized trial. *Fertility and Sterility*, 2004 81, 551–555.

Geijsen, N; Horoschak, M; Kim, K; Gribnau, J; Eggan, K; Daley, GQ. Derivation of embryonic germ cells and male gametes from embryonic stem cells. *Nature*, 2004 427,148-54.

Genbacev, O; Krtolica, A; Zdravkovic, T; Brunette, E; Powell,S; Nath, A; Caceres, E; McMaster, M; McDonagh, S; Li,Y; Mandalam, R; Lebkowski, J; Fisher, SJ. Serum-free derivation of human embryonic stem cell lines on human placental fibroblast feeders. *Fertil Steril*, 2005 83, 1517-29.

Geschwind, DH; Ou, J; Easterday, MC; Dougherty, JD; Jackson, RL; Chen, Z; Antoine, H; Terskikh, A; Weissman, IL; Nelson, SF; Kornblum HI. A genetic analysis of neural progenitor differentiation. *Neuron* 2001; 29:325–339.

Gossler, A; Doetschman, T; Korn, R; Serfling, E; Kemler, R. Transgenesis by means of blastocyst-derived embryonic stem cell lines. *Proc Natl Acad Sci USA*, 1986 83, 9065–9069.

Graf, L; Torok-Storb, B. Identification of a novel DNA sequence differentially expressed between normal human CD341CD38hi and CD341CD38lo marrow cells, *Blood*,1995 86, 548–556.

Gratsch, TE; O'Shea, KS. Noggin and chordin have distinct activities in promoting lineage commitment of mouse embryonic stem (ES) cells. *Dev Biol*, 2002 , 245, 83-94.

Greene, LA; Tischler, AS. Establishment of a noradrenergic clonal line of rat adrenal pheochromocytoma cells which respond to nerve growth factor. *Proc Natl Acad Sci USA*, 1976 73, 2424-8.

Guo, Y; Costa, R; Ramsey, H; Starnes, T; Vance, G; Robertson, K; Kelley, M; Reinbold, R; Scholer, H; Hromas, R. The embryonic stem cell transcription factors Oct-4 and FoxD3 interact to regulate endodermal-specific promoter expression. *Proc Natl Acad Sci USA*, 2002 99, 663–667.

Hanna, LA; Foreman, RK; Tarasenko, IA; Kessler, DS; Labosky, PA. Requirement for Foxd3 in maintaining pluripotent cells of the early mouse embryo. *Genes Dev*, 2002 16, 2650–2661.

Hansson, M; Tonning, A; Frandsen, U; Petri, A; Rajagopal, J; Englund, MC; Heller, RS; Hakansson, J; Fleckner, J; Skold, HN; Melton, D; Semb, H; Serup, P. Artifactual insulin release from differentiated embryonic stem cells. *Diabetes*, 2004 53, 2603-9.

Hardarson, T; Caisander, G; Sjogren, A; Hanson, C; Hamberger, L; Lundin, K. A morphological and chromosomal study of blastocysts developing from morphologically suboptimal human pre-embryos compared with control blastocysts. *Hum Reprod*, 2003 18, 399-407.

Hardy, RR; Malissen, B. Lymphocyte development .The (knock-) ins and outs of lymphoid development. *Curr Opin Immunol*, 1998 10, 155-157.

Harlan, DM; Kirk, AD. The future of organ and tissue transplantation: can T-cell co-stimulatory pathway modi- fiers revolutionize the prevention of graft rejection?. *JAMA*, 1999 282, 1076.

Heinrich, PC; Behrmann, I; Muller-Newen, G; Schaper, F; Graeve, L. Interleukin-6-type cytokine signalling through the gp130/jak/stat pathway, *Biochem J*, 1998 334 (Pt 2), 297-314.

Heins, N; Englund, MC; Sjoblom, C; Dahl, U; Tonning, A; Bergh,C; Lindahl,A; Hanson,C; Semb,H. Derivation, characterization, and differentiation of human embryonic stem cells. *Stem Cells*, 2004 22, 367-376.

Henderson, JK; Draper, JS; Baillie, HS; Fishel, S; Thomson, JA; Moore, H; Andrews, PW. Pre implantation human embryos and embryonic stem cells show comparable expression of stage specific embryonic antigens. *Stem Cells*, 2002 20, 329–337.

Hescheler, J; Fleischmann, BK; Lentini, S; Maltsev, VA; Rohwedel, J; Wobus, AM; Addicks, K. Embryonic stem cells: a model to study structural and functional properties in cardiomyogenesis. *Cardiovasc Res*, 1997 36, 149–162.

Hochedlinger, K; Jaenisch, R. Nuclear transplantation,embryonic stem cells, and the potential for cell therapy. *N Engl J Med*, 2003 349, 275-286.

Hochedlinger, K; Rideout, WM; Kyba, M; Daley, GQ; Blelloch, R; Jaenisch, R. Nuclear transplantation, embryonic stem cells and the potential for cell therapy. *Hematol J*, 2004 5, S114-S117.

Hogan, B; Constantini, F; Lacy, E. Manipulating the mouse embryo: a laboratory manual. *Cold Spring Harbor*, NY: Cold Spring Harbor Laboratory, 1994, 254–290.

Hong-mei, P; Gui-an, C. Serum-free medium cultivation to improve efficacy in establishment of human embryonic stem cell lines. Hum Reprod. 2006 21, 217-222.

Hong, Y; Winkler, C; Schartl, M. Production of medakafish chimerasfrom a stable embryonic stem cell line. *Proc Natl Acad Sci USA*, 1998 95, 3679-3684.

Hooper, M; Hardy, K; Handyside, A; Hunter, S; Monk, M. HPRT-deficient (Lesch-Nyhan) mouse embryos derived from germ line colonization by cultured cells. *Nature*, 1987 326, 292–295.

Horger, BA; Nishimura, MC; Armanini, MP; Wang, LC; Poulsen, KT; Rosenblad, C; Kirik, D; Moffat, B; Simmons, L; Johnson, E; Milbrandt, J; Rosenthal, A; Bjorklund, A;

Vandlen, RA; Hynes, MA; Phillips, HS. Neurturin exerts potent actions on survival and function of midbrain dopaminergic neurons. *J Neurosci*, 1998 18, 4929-37.

Hori, Y; Rulifson, IC; Tsai, BC; Heit, JJ; Cahoy, JD; Kim, SK. Growth inhibitors promote differentiation of insulin-producing tissue from embryonic stem cells. *Proc Natl Acad Sci USA,* 2002 99, 16105–10.

Horwitz, EM; Prockop, DJ; Fitzpatrick, LA; Koo, WW; Gordon, PL; Neel, M; Sussman M; Orchard, P; Marx, JC; Pyeritz, RE; Brenner, MK. Transplantability and therapeutic effects of bone marrow derived mesenchymal cells in children with osteogenesis imperfecta. *Nat Med,* 1999 5,309-13.

Hosler, BA; Rogers, MB; Kozak, CA; Gudas, LJ. An octamer motif contributes to the expression of the retinoic acid-regulated zinc finger gene Rex-1 (Zfp-42) in F9 teratocarcinoma cells. *Mol Cell Biol,* 1993 13, 2919–2928.

Houard, N; Rousseau, GG; Lemaigre, FP. HNF-6-independent differentiation of mouse embryonic stem cells into insulin-producing cells. *Diabetologia*, 2003 46,378–85.

Hovatta, O; Mikkola, M; Gertow, K; Stromberg, AM; Inzunza, J; Hreinsson, J; Rozell, B; Blennow, E; Andang, M; Ahrlund-Richter, L. A culture system using human foreskin fibroblasts as feeder cells allows production of human embryonic stem cells. *Hum Reprod,* 2003 18, 1404-9.

Hubner, K; Fuhrmann, G; Christenson, LK; Kehler, J; Reinbold, R; De La Fuente, R; Wood, J; Strauss, JF; Boiani, M; Scholer, HR. Derivation of oocytes from mouse embryonic stem cells. *Science*, 2003 300, 1251-6.

Hurlbut, W. *Monitoring Stem Cell Research* 53–108 (President's Council on Bioethics, Washington DC, 2004).

Humphrey, RK; Beattie, GM; Lopez, AD; Bucay, N; King, CC; Firpo, MT; Rose-John, S; Hayek, A. Maintenance of pluripotency in human embryonic stem cells is STAT3 independent. *Stem Cells*, 2004 22, 522-30.

Iannaccone, PM; Taborn, GU; Garton, RL; Caplice, MD; Brenin, DR. Pluripotent embryonic stem cells from the rat are capable of producing chimeras. *Dev Biol*, 1994 163, 288–292.

Inzunza, J; Gertow, K; Stromberg, MA; Matilainen, E; Blennow, E; Skottman, H; Wolbank, S; Ahrlund-Richter, L; Hovatta, O. Derivation of human embryonic stem cell lines in serum replacement medium using postnatal human fibroblasts as feeder cells. *Stem Cells,* 2005 23, 544-9.

Inzunza, J; Sahlen, S; Holmberg, K; Stromberg, AM; Teerijoki, H; Blennow, E; Hovatta, O; Malmgren, H .Comparative genomic hybridization and karyotyping of human embryonic stem cells reveals the occurrence of an isodicentric X chromosome after long - term cultivation .*Molecular Human Reproduction*, 2004 10,461–466.

Itskovitz-Eldor, J; Schuldiner, M; Karsenti, D; Eden, A; Yanuka, O; Amit, M; Soreq, H; Benvenisty, N. Differentiation of human embryonic stem cells into embryoid bodies compromising the three embryonic germ layers. *Mol Med*, 2000 6, 88-95.

Ivanova, NB; Dimos, JT; Schaniel, C; Hackney, JA; Moore, KA; Lemischka, IR. A stem cell molecular signature. *Science,* 2002 298, 601–604.

Jaffredo, T; Chestier, A; Bachnou, N; Dieterlen-Lievre, F. MC29-immortalized clonal avian heart cell lines can partially differentiate in vitro. *Exp Cell Res*, 1991 192, 481-491.

Jiang, Y; Jahagirdar, BN; Reinhardt, RL; Schwartz, RE; Keene, CD; Ortiz-Gonzalez, XR; Reyes, M; Lenvik, T; Lund, T; Blackstad, M; Du, J; Aldrich, S; Lisberg, A; Low, WC; Largaespada, DA; Verfaillie, CM. Pluripotency of mesenchymal stem cells derived from adult marrow. *Nature*, 2002 418, 41–49.

Kawasaki, H; Mizuseki, K; Nishikawa, S; Kaneko, S; Kuwana, Y; Nakanishi, S; Nishikawa, SI; Sasai, Y. Induction of midbrain dopaminergic neurons from ES cells by stromal cell-derived inducing activity. *Neuron*, 2000 28, 31–40.

Kawasaki, H; Suemori, H; Mizuseki, K; Watanabe, K; Urano, F; Ichinose, H; Haruta, M; Takahashi, M; Yoshikawa, K; Nishikawa, SI; Nakatsuji, N; Sasai, Y. Generation of dopaminergic neurons and pigmented epithelia from primate ES cells by stromal cell - derived inducing activity. *Proc Natl Acad Sci USA*, 2002 99, 1580– 1585.

Kawase, E; Suemori, H; Takahashi, N; Okazaki, K; Hashimoto, K; Nakatsuji, N. Strain difference in establishment of mouse embryonic stem (ES) cell lines. *Int J Dev Biol,* 1994 38, 385–390.

Kehat, I; Kenyagin-Karsenti, D; Snir, M; Segev, H; Amit, M; Gepstein, A; Livne, E; Binah, O; Itskovitz-Eldor, J; Gepstein, L. Human embryonic stem cells can differentiate into myocytes with structural and functional properties of cardiomyocytes. *J Clin Invest*, 2001 108, 407–414.

Kehat, I; Gepstein, A; Spira, A; Itskovitz-Eldor, J; Gepstein, L. High-resolution electrophysiological assessment of human embryonic cell-derived cardiomyocytes: a novel in vitro model for the study of conduction. *Circ Res*, 2002 91, 659–661.

Kelly, DL; Rizzino, A. DNA microarray analyses of genes regulated during the differentiation of embryonic stem cells. *Mol Reprod Dev*, 2000 56,113–123.

Kim, JH; Auerbach, JM; Rodriguez-Gomez, JA; Velasco, I; Gavin, D; Lumelsky, N; Lee, SH; Nguyen, J; Sanchez-Pernaute, R; Bankiewicz, K; McKay, R. Dopamine neurons derived from embryonic stem cells function in an animal model of Parkinson's disease. *Nature* 2002 418, 50-6.

Kim,SJ; Lee, JE; Park,JH; Lee,JB; Kim, JM; Yoon,BS; Song,JM; Roh,SI; Kim,CG; Yoon, HS. Efficient derivation of new human embryonic stem cell lines. *Mol Cells*, 2005 19, 46-53.

Kim, SK; Hebrok, M; Melton, DA. Notochord to endoderm signaling is required for pancreas development. *Development*, 1997 124, 4243-52.

Kimes, BW; Brandt, BL. Properties of a clonal muscle cell line from rat heart. *Exp Cell Res*, 1976 98, 367-381.

Kleinsmith, LJ; Pierce, GB. Multi potentiality of single embryonal carcinoma cells. *Cancer Res,* 1964 24, 1544–52

Klimanskaya, I; Chung, Y; Meisner, L; Johnson, J; West, MD; Lanza, R. Human embryonic stem cells derived without feeder cells. *Lancet*, 2005 365, 1636-41.

Klug, MG; Soonpaa, MH; Koh, GY; Field, LJ. Genetically selected Cardiomyocytes from differentiating embryonic stem cells form transstable intracardiac grafts. *J Clin Invest,* 1996 98, 216–224.

Kobayashi, N; Miyazaki, M; Fukaya, K; Inoue, Y; Sakaguchi, M; Uemura, T; Noguchi, H; Kondo, A; Tanaka, N; Namba, M. Transplantation of highly differentiated immortalized human hepatocytes to treat acute liver failure. *Transplantation*, 2000 69, 202-7.

Koller, BH; Smithies, O. Altering genes in animals by gene targeting *Annu Rev Immunol,* 1992 10,705– 730.

Kolossov, E; Fleischmann, BK; Liu, Q; Bloch, W; Viatchenko-Karpinski, S; Manzke, O; Ji, GJ; Bohlen, H; Addicks, K; Hescheler, J. Functional characteristics of ES cell-derived cardiac precursor cells identified by tissue- specific expression of the green fluorescent protein. *J Cell Biol,* 1998 143,2045–2056.

Korchynskyi, O; Ten Dijke, P. Identification and functional characterization of distinct critically important bone morphogenetic protein-specific response elements in the Id1 promoter. *J Biol Chem*, 2002 277, 4883–4891.

Krause, DS; Theise, ND; Collector, MI; Henegariu, O; Hwang, S; Gardner, R; Neutzel, S; Sharkis SJ. Multi-organ, multi-lineage engraftment by a single bone marrow-derived stem cell. *Cell,* 2001 105,369–377.

Krieglstein, K; Henheik, P; Farkas, L; Jaszai, J; Galter, D; Krohn, K; Unsicker, K. Glial cell line-derived neurotrophic factor requires transforming growth factor-for exerting its full neurotrophic potential on peripheral and CNS neurons. *J Neurosci*, 1998 18, 9822-34.

Kriz, V; Anneren, C; Lai C, Karlsson J, Mares J, Welsh M. The SHB adapter protein is required for efficient multilineage differentiation of mouse embryonic stem cells. *Exp Cell Res* 2003; 286:40–56.

Ku, HT; Zhang, N; Kubo, A; O'Connor, R; Mao, M; Keller, G; Bromberg, JS. Committing embryonic stem cells to early endocrine pancreas in vitro. *Stem Cells*. 2004 22,1205-17.

Labosky, PA; Barlow, DP; Hogan, BL. Mouse embryonic germ (EG) cell lines: transmission through the germ line and differences in the methylation imprint of insulin-like growth factor 2 receptor (Igf2r) gene compared with embryonic stem (ES) cell lines. *Development,* 1994 120, 3197-204.

Lacham-Kaplan O. In vivo and in vitro differentiation of male germ cells in the mouse. *Reproduction*. 2004 128, 147-52.

Lanzendorf, SE; Boyd, CA; Wright, DL; Muasher, S; Oehninger, S; Hodgen, GD. Use of human gametes obtained from anonymous donors for the production of human embryonic stem cell lines. *Fertil Steril*,2001 76,132–7.

Lederman, B; Burki, K. Establishment of a germ-line component C57BL/6 embryonic stem cell line.*Exp Cell Res,* 1991 197, 254–258.

Lee, JB; Lee, JE; Park, JH; Kim, SJ; Kim, MK; Roh, SI; Yoon, HS. Establishment and maintenance of human embryonic stem cell lines on human feeder cells derived from uterine endometrium under serum-free condition. *Biol Reprod,* 2005 72, 42-9.

Lee, JB; Song, JM; Lee, JE; Park, JH; Kim, SJ; Kang, SM; Kwon, JN; Kim, MK; Roh, SI; Yoon, HS. Available human feeder cells for the maintenance of human embryonic stem cells. *Reproduction,* 2004 128, 727-35.

Lee, SH; Lumelsky, N; Studer, L; Auerbach, JM; McKay, RD. Efficient generation of midbrain and hindbrain neurons from mouse embryonic stem cells. *Nat Biotechnol,* 2000 18, 675-9.

Lendahl, U; Zimmerman, LB; Mckay, RD. CNS stem cells express a new class of intermediate filament protein. *Cell*, 1990 60, 585-95.

Leon-Quinto, T; Jones, J; Skoudy, A; Burcin, M; Soria, B. In vitro directed differentiation of mouse embryonic stem cells into insulin-producing cells. *Diabetologia*, 2004 47, 1442-51.

Lester, LB; Kuo, HC; Andrews, L; Nauert, B; Wolf, DP. Directed differentiation of rhesus monkey ES cells into pancreatic cell phenotypes. *Reprod Biol Endocrinol*, 2004 2, 1-5.

Li, A; Simmons, PJ; Kaur; P. Identification and isolation of candidate human keratinocyte stem cells based on cell surface phenotype. *Proc Natl Acad Sci USA*, 1998a 95, 3902-7.

Li, H; Arber, S; Jessel, TM; Edlund, H. Selective agenesis of the dorsal pancreas in mice lacking homeobox gene Hlxb9. *Nat Genet*, 1999 23, 67-70.

Li, M; Pevny, L; Lovell-Badge, R; Smith, A. Generation of purified neural precursors from embryonic stem cells by lineage selection. *Curr Biol*, 1998b 8, 971-4.

Li, M; Zhang, D; Hou, Y; Jiao, L; Zhang, X; Wang, WH. Isolation and culture of embryonic stem cells from porcine blastocysts. *Mol Reprod Dev*, 2003 65, 429-434.

Li, RK; Weisel, RD; Mickle, DA; Jia, ZQ; Kim, EJ; Sakai, T; Tomita, S; Schwartz, L; Iwanochko, M; Husain, M; Cusimano, RJ; Burns, RJ; Yau, TM. Autologous porcine heart cell transplantation improved heart function after a myocardial infarction. *J Thorac Cardiovasc Surg*, 2000 119, 62-8.

Li, XJ; Du, ZW; Zarnowska, ED; Pankratz, M; Hansen, LO; Pearce, RA; Zhang, SC. Specification of motoneurons from human embryonic stem cells. *Nat Biotechnol*, 2005 23, 215-21.

Lim, JW; Bodnar, A. Proteome analysis of conditioned medium from mouse embryonic fibroblast feeder layers which support the growth of human embryonic stem cells. *Proteomics*,2002 2,1187-203.

Ling, ZD; Potter, ED; Lipton, JW; Carvey, PM. Differentiation of mesencephalic progenitor cells into dopaminergic neurons by cytokines. *Exp Neurol*, 1998 149, 411-23.

Lopez-Rovira, T; Chalaux, E; Massague, J; Rosa, JL; Ventura, F. Direct binding of Smad1 and Smad4 to two distinct motifs mediates bone morphogenetic protein-specific transcriptional activation of Id1 gene. *J Biol Chem*, 2002 277, 3176-85.

Loring, JF; Porter, JG; Seilhammer, J; Kaser, MR; Wesselschmidt, R. A gene expression profile of embryonic stem cells and embryonic stem cell-derived neurons. *Restor Neurol Neurosci,* 2001 18, 81–88.

Ludwig, TE; Levenstein, ME; Jones, JM; Berggren, WT; Mitchen, ER; Frane, JL; Crandall, LJ; Daigh, CA; Conard, KR; Piekarczyk, MS;, Llanas, RA; Thomson, JA. Derivation of human embryonic stem cells in defined conditions. Nat. Biotechnol. 2005, 24, 1-3.

Lumelsky, N; Blondel, O; Laeng, P; Velasco, I; Ravin, R; McKay, R. Differentiation of embryonic stem cells to insulin-secreting structures similar to pancreatic islets. *Science*, 2001 292,1389–94.

Luo, Y; Cai, J; Ginis, I; Sun, Y; Lee, S; Yu, SX; Hoke, A; Rao, M. Designing,testing, and validating a focused stem cell microarray for characterization of neural stem cells and progenitor cells. *Stem Cells*, 2003 21, 575–587.

Luskin, MB. Neuronal cell lineage in the vertebrate central nervous system. *FASEB J*, 1994 8, 722-30.

Maltsev, VA; Ji, GJ; Wobus, AM; Fleischmann, BK; Hescheler, J. Establishment of beta-adrenergic modulation of L-type Ca^{2+} current in the early stages of cardiomyocyte development. *Circ Res*, 1999 84, 136-145.

Manasek, FJ. Histogenesis of the embryonic myocardium. *Am J Cardiol,* 1970 25, 149–168.

Mann, JR. Imprinting in the germ line. *Stem Cells*, 2001 19: 287–294.

Marshak, DR; Gottlieb, D; Kiger, AA; Fuller, MT; Kunath, T; Hogan, B; Gardner, RL; Smith, A; Klar, AJS; Henrique, D; D'Urso, G; Datta, S; Holliday, R; Astle, C.M; Chen, J; Harrison, DE; Xie, T; Spradling, A; Andrews, PW; Przyborski, SA; Thomson, JA; Kunath, T; Strumpf, D; Rossant, J; Tanaka, S; Orkin, SH; Melchers, F; Rolink, A; Keller, G; Pittenger, MF; Marshak, DR; Flake, AW; Panicker, MM; Rao, M; Watt, FM; Grompe, M; Finegold, MJ; Kritzik, MR; Sarvetnick, N; and Winton, D.J. (2001). *Stem cell biology,* Marshak, D.R., Gardner, R.L., and Gottlieb, D. eds. (Cold Spring Harbor, New York: Cold Spring Harbor Laboratory Press).

Martin, G. Isolation of a pluripotent cell line from early mouse embryos cultured in medium conditioned by teratocarcinoma stem cells. *Proc. Natl. Acad. Sci. USA* 1981 78, 7634–7638.

Martin, GR; Evans, MJ. The formation of embryoid bodies in vitro by homogeneous embryonal carcinoma cell cultures derived from isolated single cells. *In: Sherman, MI; Solter, D .Teratomas and Differentiation.* New York: Academic; 1975, 169–87

Martin, GR. Teratocarcinomas and mammalian embryogenesis. *Science,*1980 209, 768–76.

Martin, MJ; Muotri,A; Gage,F; Varki,A. Human embryonic stem cells express an immunogenic nonhuman sialic acid. *Nat Med*, 2005 11, 228-32.

Mateizel, I; De Temmerman, N; Ullmann, U; Cauffman, G; Sermon, K; Van de Velde, H; De Rycke, M; Degreef, E; Devroey, P; Liebaers, I; Van Steirteghem, A. Derivation of human embryonic stem cell lines from embryos obtained after IVF and after PGD for monogenic disorders. *Hum Reprod*. 2006 21, 503-511.

Matsui, Y; Zsebo, K; Hogan, BLM. Derivation of pluripotential embryonic stem cells from murine primordial germ cells in culture. *Cell*, 1992 70, 841–47.

McDonald, JW; Liu, XZ; Qu, Y; Liu, S; Mickey, SK; Turetsky, D; Gottlieb, DI, Choi, DW. Transplanted embryonic stem cells survive, differentiate and promote recovery in injured rat spinal cord. *Nat Med*,1999 5,1410-2.

McKay, RD. The origins of cellular diversity in the mammalian central nervous system. *Cell,*1989 58, 815-21.

McMahon, AP; Bradley, A. The Wnt-1 (int-1) proto-oncogene is required for development of a large region of the mouse brain. *Cell*, 1990 62, 1073–1085.

Meissner, A; Jaenisch R. Generation of nuclear transfer-derived pluripotent ES cells from cloned Cdx2-deficient blastocysts. Nature. 2006 439,212-215.

Min, JY; Yang, Y; Converso, KL; Liu, L; Huang, Q; Morgan, JP; Xiao, YF. Transplantation of embryonic stem cells improves cardiac function in post infarcted rats. *J Appl Physiol*, 2002 92, 288–296.

Mitalipova, M, Calhoun, J; Shin, S; Wininger, D; Schulz, T; Noggle, S; Venable, A; Lyons, I; Robins, A; Stice, S. Human embryonic stem cell lines derived from discarded embryos. *Stem Cells,* 2003 21, 521-6.

Mitalipova, MM; Rao, RR; Hoyer, DM; Johnson, JA; Meisner, LF; Jones, KL; Dalton, S; Stice, SL. Preserving the genetic integrity of human embryonic stem cells. *Nat Biotechnol*, 2005 23, 19-20.

Mitsui, K; Tokuzawa, Y; Itoh, H; Segawa, K; Murakami, M; Takahashi, K; Maruyama, M; Maeda, M; Yamanaka, S. The homeoprotein nanog is required for maintenance of pluripotency in mouse epiblast and ES cells. *Cell*, 2003 113, 631–642.

Miyamoto, K; Hayashi, K; Suzuki, T; Ichihara, S; Yamada, T; Kano, Y; Yamabe, T; Ito, Y. Human placenta feeder layers support undifferentiated growth of primate embryonic stem cells. *Stem Cells*, 2004 22, 433– 40.

Miyazaki, S; Yamato, E; Miyazaki, J. Regulated expression of pdx-1 promotes in vitro differentiation of insulin-producing cells from embryonic stem cells. *Diabetes*,2004 53, 1030-7.

Molofsky, AV; Pardal, R; Morrison, SJ. Diverse mechanisms regulate stem cell self-renewal. *Curr Opin Cell Biol*, 2004 16, 700-7.

Moritoh, Y; Yamato, E; Yasui, Y; Miyazaki, S; Miyazaki, J. Analysis of insulin-producing cells during in vitro differentiation from feeder-free embryonic stem cells. *Diabetes*, 2003 52, 1163-8.

Movassat, J; Beattie Gillian, M; Lopez Ana, D; Hayek, A. Exendin 4 up-regulates expression of PDX 1 and hastens differentiation and maturation of human fetal pancreatic cells. *J Clin Endocrinol Metab,* 2002 87, 4775-81.

Muller, M; Fleischmann, BK; Selbert, S; Ji, GJ; Endl, E; Middeler, G; Muller, OJ; Schlenke, P; Frese, S; Wobus, AM; Hescheler, J; Katus, HA; Franz,WM. Selection of ventricular-like cardiomyocytes from ES cells in vitro. *FASEB J,* 2000 14, 2540– 2548.

Mummery, C; Oostwaard, DW; Doevendans, P; Spijker, R; Van Den Brink, S; Hassink, R; Van Der Heyden, M; Opthof, T; Pera, M; De La Riviere, AB; Passier, R; Tertoolen, L. Differentiation of human embryonic stem cells to cardiomyocytes: role of coculture with visceral endoderm-like cells. *Circulation*, 2003 107, 2733.

Nagy, A; Gocza,E; Diaz,EM; Prideaux,VR; Ivanyi,E; Markkula,M; Rossant,J. Embryonic stem cells alone are able to support fetal development in the mouse. *Development*,1990 110, 815-21.

Nagy, A; Rossant, J; Nagy, R; Abramov-Newerly, W; Roder, JC.Derivation of completely cell culture-derived mice from early-passage embryonic stem cells. *Proc Natl Acad Sci USA,* 1993 90, 8424–8428.

Nichols, J; Chambers,I; Taga,T; Smith, A: Physiological rationale for responsiveness of mouse embryonic stem cells to gp130 cytokines. *Development*, 2001 128, 2333-2339.

Nichols, J; Zevnik,B; Anastassiadis,K; Niwa,H; Klewe-Nebenius,D; Chambers,I; Scholer,H; Smith A. Formation of pluripotent stem cells in the mammalian embryo depends on the POU transcription factor Oct4. *Cell*, 1998 95, 379–391.

Niwa, H; Burdon, T; Chambers, I; Smith, A. Self-renewal of pluripotent embryonic stem cells is mediated via activation of STAT3. *Genes Dev*, 1998 12, 2048-2060.

Niwa, H; Miyazaki, J; Smith, AG. Quantitative expression of Oct-3/4 defines differentiation, dedifferentiation or self-renewal of ES cells. *Nat Genet*, 2000 24, 372–376.

Odorico, JS; Kaufman, DS; Thomson, JA. Multilineage differentiation from human embryonic stem cell lines. *Stem Cells*, 2001 19,193-204.

Oh, SK; Kim, HS; Ahn, HJ; Seol, HW; Kim, YY; Park, YB; Yoon, CJ; Kim, DW; Kim, SH; Moon, SY. Derivation and characterization of new human embryonic stem cell lines: SNUhES1, SNUhES2, and SNUhES3. *Stem Cells*, 2005a 23, 211-9.

Oh, SK; Kim, HS; Park, YB; Seol, HW; Kim, YY; Cho, MS; Ku, SY; Choi, YM; Kim, DW; Moon,SY.Methods for expansion of human embryonic stem cells. *Stem Cells*, 2005b 23, 605-9.

Okabe, S; Forsberg-Nilsson, K; Spiro, AC; Segal, M; McKay,RD. Development of neuronal precursor cells and functional post mitotic neurons from embryonic stem cells in vitro. *Mech Dev*, 1996 59, 89-102.

Okano, H. Neural stem cells: progression of basic research and perspective for clinical application. *Keio J Med*, 2002 51, 115–128.

Okuda, A; Fukushima, A; Nishimoto, M; Orimo, A; Yamagishi, T; Nabeshima, Y; Kuro-o, M; Nabeshima Yi Boon, K; Keaveney, M; Stunnenberg, H; Muramatsu, M. UTF1, a novel transcriptional coactivator expressed in pluripotent embryonic stem cells and extra-embryonic cells. *EMBO J,* 1998 17, 2019–2032.

Otonkoski, T; Beattie, GM; Mally, MI; Ricordi, C; Hayek, A. Nicotinamide is a potent inducer of endocrine differentiation in cultured human fetal pancreatic cells. *J Clin Invest*, 1993 92, 1459-66.

Pain, B; Clark, ME; Shen, M; Nakazawa, H; Sakurai, M; Samarut, J; Etches, RJ. Long-term in vitro culture and characterisation of avian embryonic stem cells with multiple morphogenetic potentialities. *Development*, 1996 122, 2339–48.

Park, IK; He, Y; Lin, F; Laerum,OD; Tian, Q; Bumgarner, R; Klug, CA; Li, K; Kuhr, C; Doyle, MJ; Xie, T; Schummer, M; Sun, Y; Goldsmith, A; Clarke, MF; Weissman, IL; Hood, L; Li, L. Differential gene expression profiling of adult murine hematopoietic stem cells. *Blood,* 2002 99, 488–498.

Park, JH; Kim, SJ; Oh, EJ; Moon, SY; Roh, SI; Kim, CG; Yoon, HS. Establishment and maintenance of human embryonic stem cells on STO, a permanently growing cell line. *Biology of Reproduction*, 2003 69, 2007–14.

Park, SP; Lee, YJ; Lee, KS; Ah Shin, H; Cho, HY; Chung, KS; Kim, EY; Lim, JH. Establishment of human embryonic stem cell lines from frozen-thawed blastocysts using STO cell feeder layers. *Hum Reprod*, 2004 19,676– 84.

Pau, KY; Wolf, DP. Derivation and characterization of monkey embryonic stem cells. *Reprod Biol Endocrinol,* 2004 2, 41.

Pebay, A; Wong, RC; Pitson, SM; Wolvetang, EJ; Peh, GS; Filipczyk, A; Koh, KL; Tellis, I; Nguyen, LT; Pera, MF. Essential roles of sphingosine-1-phosphate and platelet-derived growth factor in the maintenance of human embryonic stem cells. Stem Cells. 2005 23, 1541-1548.

Pedersen, RA.Studies of in vitro differentiation with embryonic stem cells. *Reprod Fertil Dev*, 1994 6, 543–552.

Pera, MF; Andrade, J; Houssami, S; Reubinoff, B; Trounson, A; Stanley, EG; Ward-van Oostwaard, D; Mummery, C. Regulation of human embryonic stem cell differentiation by BMP-2 and its antagonist noggin. *J Cell Sci*, 2004 117(Pt 7), 1269-80.

Pera, MF; Cooper, S; Mills, J; Parrington, JM. Isolation and characterization of a multipotent clone of human embryonal carcinoma cells. *Differentiation*, 1989 42, 10–23.

Pera, MF; Reubinoff, B; Trounson, A. Human embryonic stem cells. *Journal of Cell Science,*2000 113, 5–10.

Pesce, M; Anastassiadis, K; and Scholer, HR. Oct-4: lessons of totipotency from embryonic stem cells. *Cells Tissues Organs*, 1999 165,144–152.

Pfeiffer, SE; Jakob, H; Mikoshiba, K; Dubois, P; Guenet, JL; Nicolas, JF; Gaillard, J; Chevance, G; Jacob, F. Differentiation of a teratocarcinoma line: preferential development of cholinergic neurons. *J Cell Biol*, 1981 88, 57-66.

Phillips, RL; Ernst, RE; Brunk, B; Ivanova, N; Mahan, MA; Deanehan, JK; Moore, KA; Overton, GC; Lemischka, IR.The genetic program of hematopoietic stem cells. *Science,* 2000 288, 1635–1640.

Pickering, SJ; Braude, PR; Patel, M;, Burns, CJ; Trussler, J; Bolton, V; Minger, S. Preimplantation genetic diagnosis as a novel source of embryos for stem cell research. *Reprod Biomed Online*, 2003 7, 353-64.

Piedrahita, JA; Moore, K; Oetama, B; Lee, CK; Scales, N; Ramsoondar, J; Bazer, FW; Ott, T. Generation of transgenic porcine chimeras using primordial germ cell-derived colonies. *Biol Reprod*, 1998 58, 1321-9.

Potten, CS. Stem cells in gastrointestinal epithelium: numbers, characteristics and death. *Philos Trans R Soc Lond B Biol Sci*, 1998 353, 821–830.

Prowse, AB; McQuade, LR; Bryant, KJ; Van Dyk, DD; Tuch, BE; Gray, PP. A proteome analysis of conditioned media from human neonatal fibroblasts used in the maintenance of human embryonic stem cells.*Proteomics*, 2005 5, 978-89.

Rajagopal, J; Anderson, WJ; Kume, S; Martinez, OI; Melton, DA.I nsulin staining of ES cell progeny from insulin uptake. *Science,* 2003 299, 363.

Ramalho-Santos, M; Yoon, S; Matsuzaki, Y; Mulligan, RC; Melton, DA. ''Stemness'': transcriptional profiling of embryonic and adult stem cells. *Science*, 2002 298,597–600.

Ramiya,VK; Maraist, M; Arfors, KE; Schatz, DA; Peck, AB; Cornelius, JG. Reversal of insulin-dependent diabetes using islets generated in vitro from pancreatic stem cells. *Nat Med* ,2000 6, 278 –282.

Rappolee,DA;Basilico,C; Patel,Y; Werb, Z. Expression and function of FGF-4 in peri-implantation development in mouse embryos.*Development,* 1994 120, 2259–2269.

Rastan, S; Robertson, EJ. X-chromosome deletions in embryo-derived (EK) cell lines associated with lack of X-chromosome inactivation. *J Embryol Exp Morphol*, 1985 90, 379–388.

Rathjen, J; Haines, BP; Hudson, KM; Nesci, A; Dunn, S; Rathjen, PD. Directed differentiation of pluripotent cells to neural lineages: homogeneous formation and differentiation of a neurectoderm population. *Development*, 2002 129, 2649-61.

Reubinoff, BE; Itsykson, P; Turetsky, T; Pera, MF; Reinhartz, E; Itzik, A; Ben-Hur, T. Neural progenitors from human embryonic stem cells. *Nat biotechnol*, 2001 19, 1134-1140.

Reubinoff, BE; Pera, MF; Fong, CY; Trounson, A; Bongso, A. Embryonic stem cell lines from human blastocysts: somatic differentiation in vitro. *Nat Biotechnol*, 2000 18, 399–404.

Rhind, SM; Taylor, JE; De Sousa, PA; King, TJ; McGarry, M;Wilmut, I. Human cloning: can it be made safe? *Nature Review Genetics*, 2003 4, 855–864.

Richards, M; Fong, CY; Chan, WK; Wong, PC; Bongso, A. Human feeders support prolonged undifferentiated growth of human inner cell masses and embryonic stem cells. *Nat Biotechnol,* 2002 20, 933– 6.

Richards, M; Tan, S; Fong, CY; Biswas, A; Chan, WK; Bongso, A. Comparative evaluation of various human feeders for prolonged undifferentiated growth of human embryonic stem cells. *Stem Cells*, 2003 21, 546 –56.

Richards, M; Tan, SP; Tan, JH; Chan, WK; Bongso, A. The transcriptome profile of human embryonic stem cells as defined by SAGE. *Stem Cells*, 2004 22, 51–64.

Rodda, S; Sharma, S; Scherer, M; Chapman, G; Rathjen, P. CRTR-1, a developmentally regulated transcriptional repressor related to the CP2 family of transcription factors. *J Biol Chem,* 2001 276, 3324–3332.

Rogers, MB; Hosler, BA; Gudas, LJ. Specific expression of a retinoic acid-regulated, zinc-finger gene, Rex-1, in preimplantation embryos, trophoblast and spermatocytes. *Development*, 1991 113, 815– 824.

Rohwedel, J; Guan, K; Wobus, AM .Induction of cellular differentiation by retinoic acid in vitro. *Cells Tissues Organs*, 1999 165,190–202.

Rohwedel, J; Maltsev, V; Bober, E; Arnold, HH; Hescheler, J; Wobus, AM. Muscle cell differentiation of embryonic stem cells reflects myogenesis in vivo: developmentally regulated expression of myogenic determination genes and functional expression of ionic currents. *Dev Biol*, 1994 164, 87-101.

Rolletschek, A; Chang, H; Guan, K; Czyz, J; Meyer, M; Wobus, AM. Differentiation of embryonic stem cell-derived dopaminergic neurons is enhanced by survival-promoting factors. *Mech Dev*, 2001 105, 93-104.

Rosler, ES; Fisk, GJ; Ares, X; Irving, J; Miura, T; Rao, MS; Carpenter, MK. Long-term culture of human embryonic stem cells in feeder- free conditions. *Developmental Dynamics*, 2004 229, 259–274.

Sasaki, E; Hanazawa, K; Kurita, R; Akatsuka, A; Yoshizaki, T; Ishii, H; Tanioka, Y; Ohnishi, Y; Suemizu, H; Sugawara, A; Tamaoki, N; Izawa, K; Nakazaki, Y; Hamada, H; Suemori, H; Asano, S; Nakatsuji, N; Okano, H; Tani K. Establishment of novel embryonic stem cell lines derived from the common marmoset (*Callithrix jacchus*). Stem Cells. 2005 23, 1304-1313.

Sato, N; Meijer, L; Skaltsounis, L; Greengard, P; Brivanlou, AH. Maintenance of pluripotency in human and mouse embryonic stem cells through activation of Wnt signaling by a pharmacological GSK-3-specific inhibitor. *Nat Med*. 2004 10, 55-63.

Sato, N; Sanjuan, IM; Heke, M; Uchida, M; Naef, F; Brivanlou, AH. Molecular signature of human embryonic stem cells and its comparison with the mouse .*Dev Biol*, 2003 260, 404-13.

Scholer, HR; Hatzopoulos, AK; Balling, R; Suzuki, N; Gruss P. A family of octamer-specific proteins present during mouse embryogenesis: evidence for germline-specific expression of an Oct factor. *EMBO J*, 1989 8, 2543–2550.

Schoonjans, L; Albricht, GM; Li, JL; Collen, D; Moreadith, RW. Pluripotential rabbit embryonic stem (ES) cells are capable of forming overt coat colour chimeras following injection into blastocysts. *Mol Reproed Dev*, 1996 45, 439-443.

Schuldiner, M; Eiges, R; Eden, A; Yanuka, O; Itskovitz-Eldor, J; Goldstein, RS; Benvenisty, N. Induced neuronal differentiation of human embryonic stem cells. *Brain Res*, 2001 913, 201-15.

Schuldiner, M; Yanuka, O; Itskovitz-Eldor, J; Melton, DA; Benvenisty, N. From the cover: effects of eight growth factors on the differentiation of cells derived from human embryonic stem cells. *Proc Natl Acad Sci USA*, 2000 97, 11307-12.

Schwitzgebel, VM; Sceel, DW; Conners, JR; Kalamaras, J; Lee, JE; Anderson, DJ; Sussel, L; Johnson, JD; Jerman, MS. Expression of neurogenin3 reveals an islet cell precursor population in the pancreas. *Development*, 2000 127, 3533-42.

Seale, P; Asakura, A; Rudnicki, MA. The potential of muscle stem cells. *Dev Cell*, 2001 1, 333–342.

Segev, H; Fishman, B; Ziskind, A; Shulman, M; Itskovitz-Eldor, J. Differentiation of human embryonic stem cells into insulin-producing clusters. *Stem Cells*, 2004 22, 265-74.

Sen, A; Dunnmon, P; Henderson, SA; Gerard, RD; Chien, KR. Terminally differentiated neonatal rat myocardial cells proliferate and maintain specific differentiated functions following expression of SV40 large T antigen. *J Biol Chem*, 1988 263, 19132-19136.

Shamblott, MJ; Axelman, J; Littlefield, JW; Blumenthal, PD; Huggins, GR; Cui, Y; Cheng, L; Gearhart, JD. Human embryonic germ cell derivatives express a broad range of developmentally distinct markers and proliferate extensively in vitro. *Proc Natl Acad Sci USA,* 2001 98, 113-8.

Shamblott, MJ; Axelman, J; Wang, S; Bugg, EM; Littlefield, JW; Donovan, PJ; Blumenthal, PD; Huggins, GR; Gearhart, JD. Derivation of pluripotent stem cells from cultured human primordial germ cells. *Proc Natl Acad Sci USA*, 1998 95, 13726-13731.

Shi, Y; Hou, L; Tang, F; Jiang, W; Wang, P; Ding, M; Deng, H. Inducing embryonic stem cells to differentiate into pancreatic beta cells by a novel three-step approach with activin A and all-trans retinoic acid. *Stem Cells*, 2005 23, 656-62.

Shin, DM; Ahn, JI; Lee, KH; Lee, YS; Lee, YS. Ascorbic acid responsive genes during neuronal differentiation of embryonic stem cells. *Neuroreport*. 2004 15, 1959-63.

Shiroi, A; Yoshikawa, M; Yokota, H; Fukui, H; Ishizaka, S; Tatsumi, K; Takahashi, Y. Identification of insulin-producing cells derived from embryonic stem cells by zinc-chelating dithizone .*Stem Cells*, 2002 20, 284–92.

Simon, C; Escobedo, C; Valbuena, D; Genbacev, O; Galan, A; Krtolica, A; Asensi, A; Sanchez, E; Esplugues, J; Fisher, S; Pellicer, A. First derivation in Spain of human embryonic stem cell lines: use of long-term cryopreserved embryos and animal-free conditions. *Fertil Steril*. 2005 83, 246-9.

Sipione, S; Eshpeter, A; Lyon, JG; Korbutt, GS; Bleackley, RC. Insulin expressing cells from differentiated embryonic stem cells are not beta cells. *Diabetologia*, 2004 47, 499-508.

Sjogren, A; Hardarson, T; Andersson, K; Caisander, G; Lundquist, M; Wikland, M; Semb, H; Hamberger, L . Human blastocysts for the development of embryonic stem cells. *Reprod Biomed Online*. 2004 9, 326-9.

Slack, JMW. Stem cells in epithelial tissues. *Science,* 2000 117, 1431–3.

Smith, AG. Origins and properties of mouse embryonic stem cells. *Annu Rev Cell Dev Biol*. 2001 17,435-62.

Snir, M; Kehat, I; Gepstein, A; Coleman, R; Itskovitz-Eldor, J; Livne, E; Gepstein, L. Assessment of the ultrastructural and proliferative properties of human embryonic stem cell-derived cardiomyocytes *Am J Physiol Heart Circ Physiol.* 2003 285, H2355-63.

Solter, D; Skreb, N; Damjanov, I. Extra uterine growth of mouse egg cylinders results in malignant teratoma. *Nature,* 1970 227, 503–4.

Soria, B; Roche, E; Berna, G; Leon-Quinto, T; Reig, JA; Martin, F. Insulin-secreting cells derived from embryonic stem cells normalize glycemia in streptozotocin-induced diabetic mice. *Diabetes*, 2000 49,157–62.

Sperger, J M; Chen, X; Draper, JS; Antosiewicz, JE; Chon, CH; Jones, SB; Brooks, JD; Andrews, PW; Brown, PO; Thomson, JA. Gene expression patterns in human embryonic stem cells and human pluripotent germ cell tumors. *Proc Natl Acad Sci USA*, 2003 100, 13350–13355.

Stevens, LC. A new inbred subline of mice (129/terSv) with a high incidence of spontaneous congenital testicular teratomas. *J Natl Cancer Inst*, 1970 50, 235–242.

Strumpf, D; Mao, CA; Yamanaka, Y; Ralston, A; Chawengsaksophak, K; Beck, F; Rossant, J. *Cdx2* is required for correct cell fate specification and differentiation of trophectoderm in the mouse blastocyst. Development. 2005 132, 2093-2102.

Stojkovic, M; Lako, M; Stojkovic, P; Stewart, R; Przyborski, S; Armstrong, L; Evans, J; Herbert, M; Hyslop, L; Ahmad, S; Murdoch, A; Strachan, T. Derivation of human embryonic stem cells from day-8 blastocysts recovered after three-step in vitro culture. *Stem Cells*, 2004 22, 790-7.

Stojkovic, M; Stojkovic, P; Leary, C; Hall, VJ; Armstrong, L; Herbert, M; Nesbitt, M; Lako, M; Murdoch, A. Derivation of a human blastocyst after heterologous nuclear transfer to donated oocytes. *Reprod Biomed Online*, 2005a 11, 226-31.

Stojkovic, P; Lako, M; Przyborski, S; Stewart, R; Armstrong, L; Evans, J; Zhang, X; Stojkovic, M. Human-serum matrix supports undifferentiated growth of human embryonic stem cells. *Stem Cells*, 2005b 23, 895-902.

Stojkovic, P; Lako, M; Stewart, R; Przyborski, S; Armstrong, L; Evans, J; Murdoch, A; Strachan, T; Stojkovic, M. An autogeneic feeder cell system that efficiently supports growth of undifferentiated human embryonic stem cells. *Stem Cells*, 2005c 23, 306-14.

Strelchenko, N. Bovine pluripotent stem cells. *Theriogenology*, 1996 45,131-140.

Strelchenko, N; Verlinsky, O; Kukharenko, V; Verlinsky, Y. Morula-derived human embryonic stem cells. *Reprod Biomed Online*, 2004 9, 623-9.

Suemori, H; Tada, T; Torii, R; Hosoi, Y; Kobayashi, K; Imahie, H; Kondo, Y; Iritani, A; Nakatsuji, N. Establishment of embryonic stem cell lines from cynomolgus monkey blastocysts produced by IVF or ICSI. *Dev Dyn*, 2001 222, 273–279.

Sukoyan MA, Golubitsa, AN; Zhelezova, AI; Shilov, AG; Vatolin,SY; Maximovsky, LP; Andreeva, LE; McWhir, J; Pack, SD; Baybordin, SI; Kerkis, AY; Hikizilova, OL; Seov .Isolation and cultivation of blastocyst derived stem cell lines from American mink (Mustela vison). *Mol Reprod Dev*, 1992 33, 418-431.

Sun, L; Bradford, CS; Ghosh, G; Collodi, P. ES- like cell cultures derived from early zebrafish embryos.Mol Mar Biol Biotechnol,1995 4,193-199.

Suss-Toby, E; Gerecht-Nir, S; Amit, M; Manor, D; Itskovitz-Eldor, J. Derivation of a diploid human embryonic stem cell line from a mononuclear zygote. *Hum Reprod*, 2004 19, 670-5.

Sutton, J; Costa, R; Klug, M; Field, L; Xu, D; Largaespada, DA; Fletcher, CF; Jenkins, NA; Copeland, NG; Klemsz, M; Hromas, R. Genesis, a winged helix transcriptional repressor with expression restricted to embryonic stem cells. *J Biol Chem*, 1996 271, 23126–23133.

Takagi, Y; Takahashi, J; Saiki, H; Morizane, A; Hayashi, T; Kishi, Y; Fukuda, H; Okamoto, Y; Koyanagi, M; Ideguchi, M; Hayashi, H; Imazato, T; Kawasaki, H; Suemori, H; Omachi, S; Iida, H; Itoh, N; Nakatsuji, N; Sasai, Y; Hashimoto, N. Dopaminergic neurons generated from monkey embryonic stem cells function in a Parkinson primate model. *J Clin Invest,* 2005 115,102-9.

Takahashi, K; Mitsui, K; Yamanaka, S. Role of eras in promoting tumour-like properties in mouse embryonic stem cells. *Nature* 2003 423, 541-545.

Tanaka, TS; Kunath, T; Kimber, WL; Jaradat, SA; Stagg, CA; Usuda, M; Yokota, T; Niwa, H; Rossant, J; Ko, MS. Gene expression profiling of embryo-derived stem cells reveals candidate genes associated with pluripotency and lineage specificity. *Genome Res*, 2002 12, 1921–1928.

Tao, L; Zhou, CQ; Mai, QY; Zhuang, GL. Establishment of human embryonic stem cell line from gamete donors. *Chin Med J (Engl).* 2005 118, 116-22.

Terskikh, AV; Easterday, MC; Li, L; Hood, L; Kornblum, HI; Geschwind,DH; Weissman, IL. From hematopoiesis to neuropoiesis: evidence of overlapping genetic programs. *Proc Natl Acad Sci U S A,* 2001 98,7934–7939.

Theise, ND; Saxena, R; Portmann, BC; Thung, SN; Yee, H; Chiriboga, L; Kumar, A; Crawford, JM .The canals of Hering and hepatic stem cells in humans. *Hepatology,*1999 30, 1425–1433.

Thomson, JA; Itskovitz-Eldor,J; Shapiro,SS; Waknitz,MA; Swiergiel,JJ; Marshall,VS; Jones,JM. Embryonic stem cell lines derived from human blastocysts. *Science*, 1998 282, 1145-7.

Thomson, JA; Kalishman, J; Golos, TG; Durning, M; Harris, CP; Hearn, JP. Pluripotent cell lines derived from common marmoset (*Callithrix jacchus*) blastocysts. *Biol Reprod*, 1996 55, 254–259.

Thomson, JA; Kalishman, J; Golos, TG; Durning, M; Harris, CP; Becker, RA; Hearn, JP. Isolation of a primate embryonic stem cell line. *Proc Natl Acad Sci USA*,1995 92, 7844 – 7848.

Tourrel, C; Bailbe, D; Lacorne, M; Meile, MJ; Kergoat, M; Portha, B. Persistent Improvement of Type 2 Diabetes in the Goto-Kakizaki Rat Model by Expansion of the {beta}-Cell Mass During the Prediabetic Period With Glucagon-Like Peptide-1 or Exendin-4. *Diabetes* 2002 51, 1443-52.

Tourrel, C; Bailbe, D; Meile, MJ; Kergoat, M; Portha, B. Glucagon-like peptide-1 and exendin-4 stimulate beta-cell neogenesis in streptozotocin-treated newborn rats resulting in persistently improved glucose homeostasis at adult age. *Diabetes*, 2001 50, 1562-70.

Toyooka, Y; Tsunekawa, N; Akasu, R; Noce, T. Embryonic stem cells can form germ cells in vitro. *Proc Natl Acad Sci U S A*, 2003 100, 11457-62.

Tremain, N; Korkko, J; Ibberson, D; Kopen, GC; DiGirolamo, C; Phinney,DG. Micro SAGE analysis of 2,353 expressed genes in a single cell derived colony of undifferentiated human mesenchymal stem cells reveals mRNAs of multiple cell ineages. *Stem Cells*, 2001 19, 408–418.

Tropepe, V; Coles, BL; Chiasson, BJ; Horsford, DJ; Elia, AJ; McInnes, RR; Van der Kooy, D. Retinal stem cells in the adult mammalian eye. *Science*, 2000 287, 2032–2036.

Vallier, L; Alexander, M; Pedersen, RA. Activin/Nodal and FGF pathways cooperate to maintain pluripotency of human embryonic stem cells. J Cell Sci. 2005 118(Pt 19):4495-4509.

Vallier, L; Rugg-Gunn, PJ; Bouhon, IA; Andersson, FK; Sadler, AJ; Pedersen, RA. Enhancing and diminishing gene function in human embryonic stem cells. *Stem Cells*, 2004 22, 2–11.

van de Stolpe, A; van den Brink, S; van Rooijen, M; Ward-van Oostwaard, D; van Inzen, W; Slaper-Cortenbach, I; Fauser, B; van den Hout, N; Weima, S; Passier, R; Smith, N; Denning, C; Mummery, C. Human embryonic stem cells: towards therapies for cardiac disease. Derivation of a Dutch human embryonic stem cell line. Reprod Biomed Online. 2005 11, 476-485.

Vassilieva, S; Guan, K; Pich, U; Wobus, AM. Establishment of SSEA-1- and Oct-4-expressing rat embryonic stem-like cell lines and effects of cytokines of the IL-6 family on clonal growth. *Exp Cell Res*, 2000 258, 361–373.

Ventura, C; Zinellu, E; Maninchedda, E; Maioli, M. Dynorphin B is an agonist of nuclear opioid receptors coupling nuclear protein kinase C activation to the transcription of cardiogenic genes in GTR1 embryonic stem cells. *Circ Res*, 2003 92: 623–629.

Verlinsky, Y; Strelchenko, N; Kukharenko, V; Rechitsky, S; Verlinsky, O; Galat, V; Kuliev, A. Human embryonic stem cell lines with genetic disorders. *Reprod Biomed Online*, 2005 10,105-10.

Viragh, S; Challice, CE. Variations in filamentous and fibrillar organization, and associated sarcolemmal structures in cells of the normal mammalian heart. *J Ultra struct Res*, 1969 28, 321–34.

Walker, PR; Saas, P; Dietrich, PY .Role of Fas ligand (CD95L) in immune escape: the tumor cell strikes back. *J. Immunol*, 1997 158, 4521.

Wang, Q; Fang, ZF; Jin, F; Lu, Y; Gai, H; Sheng, HZ. Derivation and growing human embryonic stem cells on feeders derived from themselves. *Stem Cells*. 2005a 23, 1221-1227.

Wang, G; Zhang, H; Zhao, Y; Li, J; Cai, J; Wang, P; Meng, S; Feng, J; Miao, C; Ding, M; Li, D; Deng, H. Noggin and bFGF cooperate to maintain the pluripotency of human embryonic stem cells in the absence of feeder layers. *Biochem Biophys Res Commun*. 2005b 330, 934-42.

Watanabe, K; Kamiya, D; Nishiyama, A; Katayama, T; Nozaki, S; Kawasaki, H; Watanabe, Y; Mizuseki, K; Sasai, Y. Directed differentiation of telencephalic precursors from embryonic stem cells. *Nat Neurosci*, 2005 8, 288-96.

Watowich, SS; Wu, H; Socolovsky, M; Klingmuller, U; Constantinescu, SN; Lodish, HF; Cytokine receptor signal transduction and the control of hematopoietic cell development, *Ann Rev Cell Dev Biol*. 1996 12, 91-128.

Watt, FM; Hogan, BL.Out of Eden: stem cells and their niches. *Science,* 2000 287,1427-30.

Watt, FM. Stem cell fate and patterning in mammalian epidermis. *Curr Opin Genet Dev,* 2001 11, 410–417.

Weissman, IL. Stem cells: units of development, units of regeneration, and units in evolution. *Cell,* 2000 100,157-68.

Wells, DN; Misica, PM; Day, TAM; Tervit, HR. Production of cloned lambs from an established embryonic cell line: A comparison between in vivo and in vitro matured cytoplasts. *Biol Reprod,* 1997 57, 385-393.

West, JA; Daley, GQ. In vitro gametogenesis from embryonic stem cells. *Curr Opin Cell Biol.* 2004 16, 688-92.

Westphal, CH; Leder, P. Transposon-generated 'knock-out' and 'knock-in' gene-targeting constructs for use in mice. *Curr Biol,*1997 7, 530-533.

Wichterle, H; Lieberam, I; Porter, JA; Jessell, TM. Directed differentiation of embryonic stem cells into motor nourons. *Cell,* 2002 110, 385-97.

Wobus, AM; Grosse,R; Schoneich,J. Specific effects of nerve growth factor on the differentiation pattern of mouse embryonic stem cells in vitro. *Biomed Biochim Acta,* 1988 47, 965-73.

Wobus, AM; Guan, K; Shan, J; Wellner, MC; Rohwedel, J; Ji G; Fleischmann, B; Katus, HA; Hescheler, J; Franz, WM. Retinoic acid accelerates embryonic stem cell-derived cardiac differentiation and enhances development of ventricular cardiomyocytes. *J Mol Cell Cardiol* 1997 29, 1525–1539.

Wobus, AM; Kleppisch,T; Maltsev, V; Hescheler,J. Cardiomyocyte-like cells differentiated in vitro from embryonic carcinoma cells P19 are characterized by functional expression of adrenoceptors and Ca2+ channels. *In Vitro Cell Dev Biol Anim,* 1994 30,425-34.

Wobus, AM; Wallukat, G; Hescheler, J. Pluripotent mouse embryonic stem cells are able to differentiate into cardiomyocytes expressing chronotropic responses to adrenergic and cholinergic agents and Ca^{2+} channel blockers. *Differentiation,* 1991 48, 173–82.

Wood, SA; Pascoe,WS; Schmidt,C; Kemler, R; Evan, AJ; Allen, ND.Simple and efficient production of embryonic stem cell-embryo chimeras by coculture. *Proc Natl Acad Sci USA,* 1993 90, 4582–4585.

Wu, X; Ding, S; Ding, Q; Gray, NS; Schultz PG. Small molecules that induce cardiomyogenesis in embryonic stem cells. *J Am Chem Soc* 2004 126,1590–1591.

Xu, C; Inokuma,MS; Denham,J; Golds,K; Kundu,P; Gold, JD;Carpenter, MK. Feeder-free growth of undifferentiated human embryonic stem cells. *Nat Biotechnol,* 2001 19, 971-4.

Xu, C; Police,S; Rao,N; Carpenter, MK. Characterization and enrichment of cardiomyocytes derived from human embryonic stem cells. *Circ Res,* 2002a 91,501–508.

Xu, C; Rosler,E; Jiang,J; Lebkowski,JS; Gold,JD; O'Sullivan,C; Delavan-Boorsma,K; Mok,M; Bronstein,A; Carpenter, MK. Basic fibroblast growth factor supports undifferentiated human embryonic stem cell growth without conditioned medium. Stem Cells, 2005a 23, 315-23.

Xu, G; Stoffers, DA; Habener, JF; Bonner-Weier, S. Exendin-4 stimulated both beta cell replication and neogenesis, resulting in increased beta cell mass and improved glucose tolerance in diabetic rats. Diabetes, 1999 48, 2270-6.

Xu, RH; Chen, X; Li, DS; Li, R; Addicks, GC; Glennon, C; Zwaka, TP;Thomson,JA. BMP4 initiates human embryonic stem cell differentiation to trophoblast. Nat Biotechnol, 2002b 20, 1261–1264.

Xu, RH; Peck, RM; Li, DS; Feng, X; Ludwig, T; Thomson, JA. Basic FGF and suppression of BMP signaling sustain undifferentiated proliferation of human ES cells. Nat Methods, 2005b 2, 185-90.

Ying, QL; Nichols, J; Chambers, I; Smith, A. BMP induction of Id proteins suppresses differentiation and sustains embryonic stem cell self-renewal in collaboration with STAT3.Cell, 2003a 115,281–292.

Ying, QL; Stavridis, M; Griffiths, D; Li, M; Smith, A. Conversion of embryonic stem cells into neuroectodermal precursors in adherent monoculture. Nat Biotechnol, 2003b 21,183-186.

Yuan, H; Corbi,N; Basilico, C; Dailey,L. Developmental-specific activity of the FGF-4 enhancer requires the synergistic action of Sox2 and Oct-3. Genes and Development, 1995 9, 2635–2645.

Zhang, SC; Wernig,M; Duncan, ID; Brustle,O; Thomson, JA.In vitro differentiation of transplantable neural precursors from human embryonic stem cells. Nat Biotechnol,2001 19,1129-33.

Zhang, W; Lee, WH; Triarhou, LC. Grafted cerebellar cells in a mouse model of hereditary ataxia express IGF-I system genes and partially restore behavioral function. Nat Med, 1996 2, 65-71.

Zhou, G; Chen, J; Lee, S; Clark, T; Rowley, JD; Wang, SM .The pattern of gene expression in human CD34 (1) stem/progenitor cells. Proc Natl Acad Sci USA, 2001 98, 13966–13971.

Zulewski, H; Abraham, EJ; Gerlach, MJ; Daniel, PB; Moritz, W; Muller, B; Vallejo, M; Thomas, MK; Habener, JF. Multipotential nestin-positive stem cells isolated from adult pancreatic islets differentiate ex vivo into pancreatic endocrine, exocrine, and hepatic phenotypes. Diabetes, 2001 50, 521-33.

Zwaka, TP; Thomson, JA.Homologous recombination in human embryonic stem cells. Nature Biotechnology, 2003 21, 319–321.

In: Frontiers in Stem Cell Research
Editor: Julia M. Spanning pp. 119-134

ISBN 1-60021-294-8

Chapter IV

Pain Care and Stem Cells: The Effect of Political Ideology on the Patient Centered Paradigm of Health Care

Anthony J. Eksterowicz[1,*] *and Nancy B. Eksterowicz*[2]

[1]James Madison University, Department of Political Science, Harrisonburg VA 22807;
[2]Advanced Practice Nurse Pain Management, University of Virginia Health System Pain Services, Charlottesville, VA 22980, USA

Abstract

This paper examines two public policy decisions and their subsequent effect upon health care. The first decision was the proposed Lethal Drug Abuse prevention Act of 1998 otherwise known as the Hyde-Nichols bill. While the bill ultimately died in the Congress the debate it generated has allowed scholars and policymakers alike to assess its impact upon health care, specifically the healthcare associated with pain management. The bill had political and ideological motivations in that its main purpose was to nullify the state of Oregon's physician assisted suicide legislation. The subsequent debate illustrates the problem of legislating from the perspective of political ideology and not pure science. The second decision is that of President Bush regarding the usage of stem cells in basic biological and medical research. The decision to use existing stem cells has already been surpassed by the needs of scientific research. This has touched off a vigorous debate concerning the need for an expanded policy regarding the use of stem cells. Both decisions pose problems for basic research and health care. We note these potential difficulties and draw conclusions concerning the role of political ideology in the health care field. We also present suggestions for alleviating some of these difficulties.

---

* Correspondence concerning this article should be addressed to Dr. Anthony J. Eksterowicz, Professor of Political Science, James Madison University, Department of Political Science, Harrisonburg VA 22807, E-mail: EKSTERAJ@JMU.EDU; Tel: 540-568-6149.

The role of political ideology in health care decisions has become increasingly important in the past decade. As science pushes the boundaries of pain relief and life replication (cloning and stem cells) issues it confronts political, ideological and ethical issues that may have detrimental impacts upon patient care, the relief of pain and cures for diseases. For this study, we examine that portion of political ideology which effects one's beliefs about the origins and preservation of life and its effect upon health care policies, specifically patient pain care and basic research regarding stem cells. We hypothesize that ideological belief systems (based partly upon religious beliefs) can have a detrimental effect upon future patient care. While we are aware that it is difficult to distinguish between political and ideological policy entrepreneurship, we also agree with Colin Campbell (2004) that ideological entrepreneurship involves a depth of feeling that makes compromise difficult. It also involves interests tied to a president's or congressperson's political base which are particularly persistent in their policy preferences.

This study examines a number of variables associated with these two areas. First we discuss the new emerging model of patient health care from the perspective of pain management practitioners. Second, we apply the model to two case studies involving political decisions. One decision is the congressionally based Hyde – Nichols bill with implications for patient pain control and the other is President Bush's decision concerning stem cells and its effect upon basic research and ultimately patient care. Third, we derive some lessons from both experiences that might help manage some of the problems concerning patient pain care and basic research in the stem cell area.

Pain Management and the New Emerging Health Care Paradigm

Individuals in our society and culture learn through various institutions (e.g., media, medicine, family) that "pain" is a deeply individual enterprise and that suffering through the pain is merely an accepted part of life. Society and culture profoundly shape our political, ideological and economic experiences. So too, culture shapes our perception of pain. David B. Morris in his classic work, *The Culture of Pain,* suggests that individual minds and specific cultures shape the experience of pain. He notes, for example the use of sentimental cultural pain in Harriet Beecher Stowe's *Uncle Tom's Cabin* and its effect upon political change. Morris warns that we cannot merely view the concept of pain as medical. Pain is not divisible. It exists within the body but also within the mind. After the terrorist attacks on September 11, 2001, pain specialists saw a dramatic increase in pain complaints from chronic pain sufferers. Peter Staats, chief of pain medicine at John Hopkins University, noted that there is a strong mind-body connection involved in pain and cultural events (Goldstein, 2001). Confining pain to the medical realm and viewing it as a neuro-matter is misguided (Morris, 1991, passim).

Morris prefers to view pain not as something that medical science can conquer or solve like a puzzle but rather as a mystery. This is very often how patients view pain. Thus pain is subjective. For example, acute pain differs from chronic pain. Acute pain can and will disappear but chronic pain befuddles patients and doctors alike. It deeply depends upon the perception of the patient. To mitigate the problem doctors must work closely with patients (Morris, 1991 pp. 9-79). But this notion of partnership contrasts with a doctor centered or driven relationship as predominate in the last century. In the past, patients received and respected their physicians due to their education, knowledge, and their remarkable ability to heal. Traditionally, doctors would decide what the correct treatment was and present it to the patient. This had some undesirable consequences. One such consequence was the paternalistic attitude reflecting the predominant cultural value that women were more emotional and may exaggerate pain complaints. Many doctors were more likely to dismiss women's complaints of pain as hysteria or neurosis. In a study cited by Post et al. (1996 p. 350), a personal and cultural gender bias of doctors and nurses continues to exert a negative influence over medication choices for analgesia and sedation. This balance of power is fundamentally shifting. Today, many health-care recipients demand relationships that respect and protect autonomy, and enhance well-being (Post, et al.1996 p. 349). Patients recognize the limitations of their doctors as well as medical science. They desire more of a partnership when it comes to healing.

If pain is perception, if it is emotional as well as physical, if it befuddles patients and doctors alike, if it requires a partnership between patient and doctor then remediation of pain must involve a broader societal approach. Dealing with pain must involve the patient, the clinicians and a society that comprehends the integration of the mind and body. It must also involve recognition of how the cultural role of pain within a society impacts treatment. There is a political consequence to such cultural pain. For example, cultural pain afflicts the socio-economically disadvantaged. A recent study, (Anderson et. al 2000) demonstrated that physicians are more likely to underestimate pain severity in minorities and women and offer insufficient analgesics for pain associated with cancer.

As Morris notes, "All pain, especially all chronic pain, is an interdependent, inseparable, multidimensional union of the two elemental human forces that the Greeks called psyche (mind) and soma (body)" (Morris, 1991 p. 158). In the treatment of pain a war of sorts broke out between the peripheralist (the dominant) and the centralist (the challenger) paradigm. Theorists from the peripheralist paradigm suggest that pain can be traced to organ or nerve damage. In other words it is organically initiated. The centralist school believes that while some pain may be peripheral, pain is more complex. For example there exists the notion of intractable pain, which serves as a paradox to peripheral notions (Morris, 1991 p. 167). If the centralists are correct the treatment of pain will be more complex and multidisciplinary in nature. It will require greater education, specialization and interdisciplinary efforts aimed at both the body and mind. It will require an extraordinary partnership with patients and new a new attitude among physicians and health care providers. It may require societal and cultural choices concerning pain, suffering and dying. These choices will ultimately be economic and political in nature involving issues such as, stem cell research, the right to die, terminal patient pain relief and efforts to remediate pain by physicians, which may conflict with our nation's drug laws. As Rowlingson (2001) notes:

> It's quite fascinating to think about moving from a traditional medical model where doctors directed nurses and nurses delivered the care to patients that had pain. With this perspective you realize that the person with the symptoms is at the end of the treatment

> chain. The concept of patient-centered care is revolutionary, and it shouldn't be. There will be enormous concerns amongst health care workers in having to work in that kind of environment where the patient has been empowered to be the one in charge.

We stand on the cusp of a paradigmatic change in pain care. However, in this era the old paradigm is very much alive. As this paradigm begins to shift, society will be confronted with anomalies. The organic model has been increasingly unable to provide answers for these anomalies. The old paradigm has fostered many myths, which have been reported by the media and adopted by other institutions as conventional wisdoms. As Daniel B. Carr (1999) notes:

> The Popular media convey inaccurate ideas about pain and pain treatment in this country. The result is misperception on the part of the public that lead to the generation and spread of myths about pain. Some of these myths are: Pain should be endured. Pain is inevitable: analgesia cannot be provided conveniently. It is wrong if terminal patients die a little sooner because of pain-relieving medication. Good patients endure pain and do not distract physicians from treating their disease. Pain medication leads to addiction. Patients are sometimes in so much pain they desire death, and there is nothing we can do about that.
>
> These statements are all incorrect and, viewed in the light of modern knowledge, are easily refuted. Nevertheless, these myths are prevalent and adversely affecting not only patient's thinking but also medical practice. Education is the means by which these myths will finally be silenced.

While Dr. Carr emphasizes the direct impact from the media, the medical profession cannot escape blame. How many general practice physicians prescribe the lower and "safer" doses of pain killing medication because of their mistaken belief that pain medication leads to addiction (Hansen, 2000 pp. 2317-18)? Misunderstanding of pain, pain treatments and effects of medication lead physicians to fear legal penalties and disciplinary action (Shapiro, 1996 p. 363). Information about what is legally and medically acceptable for the treatment of pain is rarely conveyed to state medical boards (Hyman, 1996 p.338). Some members of the medical profession are precipitating a silent crisis of patient under-medication across this nation.

The myths Carr describes are, in part, the result of the conflict between the old peripheralist paradigms with the new centralist paradigm. The public generally accepts these myths. They have meaning and directly and indirectly affect our economics and politics with respect to the management of pain.

The conflicting nature of these two paradigms must also be placed within the context of the two competing health care paradigms. The old patient care paradigm was essentially paternalistic and is being supplanted by a new patient autonomous paradigm (Post, et al., 1996 pp. 348-49). Paradigms change slowly over time. Already we can observe some progress. Health care regulatory agencies such as Centers for Medicare and Medicaid, (CMS) formerly known as HCFA, and Joint Commission on Accreditation of Healthcare Organizations (JCAHO) are incorporating pain management standards into the evaluation and accreditation process (PM Standards JCAHO: 1999 online). This comes as a response from the public, pain management organizations and pain specialists. However, our economics and politics may be trapped within the old paradigm. Efforts at education will help change attitudes, laws and markets but not without resistance and real cultural pain. These economic impediments work against the alleviation of patient suffering and against the emerging patient centered health care paradigm. One example of the impact of political ideology upon societal patient pain care is the Hyde-Nichols bill.

The Hyde-Nichols Bill

The now defunct Hyde-Nichols bill otherwise known as the Lethal Drug Abuse Prevention Act (House Bill 2260, Senate Bill 1272) was proposed in 1998. It essentially amended the 1970 Control Substance Act to prevent physician assisted suicide. The state of Oregon had passed a physician assisted suicide statute known as the Death with Dignity Act. Both Representative Hyde (R-Ill) and Senator Nichols (R-OK) were opposed to this nascent state movement due to their strong pro-life positions. Their bill would have barred doctors from using federally regulated drugs (schedule II drugs) to aid in patient suicide. Under this proposed act any physician *intentionally* using a controlled substance for the purposes of assisting in a patient's suicide could lose their license. In addition, the act established a board of experts which would advise the Drug Enforcement Agency (DEA) in murky cases where it was not clear if the physician was assisting in patient suicide or aggressively providing pain control (Orentflicker and Caplin, 2000 pp. 255-256).

The language of the bill caused uproar in the health care community. For example, how could one determine the intent of the physician? The line between providing pain control and aiding in patient suicide is very fine. Furthermore, health care professionals feared the establishment of a board of experts presumably drawn from the health care field to advise the DEA. They believed that the DEA would play a larger role in their field. This coupled with the low levels of knowledge concerning pain relief could prove detrimental to the alleviation of pain nation wide. Both the American Medical Association and the National Hospice Organization opposed the bill (Orentlicher and Caplin, 2000 pp. 255-8).

The bill also represented the intersection of political ideology with health care and health care regulation. The bill reflected the pro-life ideological persuasion of both Representative Hyde and Senator Nichols. Legislating according to ideology and not science could have negative consequences for the alleviation of patient suffering. If ideology or religious beliefs and science conflict which path should society choose? We are now witnessing the same type of debate concerning the stem cell issue (Connolly, et. al., 2001; Will, 2001; Cohen, 2001). This issue tends to fit Campbell's (2004) idea of ideological policy entrepreneurship in that it involved a situation where compromise was difficult and significant portions of both Hyde and Nichol's persistent political base (pro-life interests) were involved in this decision.

Due to the heavy criticism of the bill it was amended and re-named in 1999. The Pain Promotion Relief Act eliminated the board of experts advising the DEA. It also included language that permitted the use of controlled substances for pain management and palliative care even if the use of such a substance may increase the risk of death. There was a section that addressed the education of law enforcement officials about the legitimate use of controlled substances for pain and palliative care. With these changes some interest groups like the American Medical Association originally endorsed the bill. However when other health care groups criticized the bill the AMA rescinded its support (Reuters Health 2000). What were the criticisms of the amended bill?

The amended bill still leaves the line between acceptable palliative care and unacceptable suicide defined by physician intent. It is difficult to determine intent. Physicians would worry that law enforcement officials could interpret intent when no intent existed. The murky line between acceptable care and end of life decisions could also affect physician intent. All of this could force doctors to be conservative and under-treat pain. Physicians utilizing aggressive drug therapies could not only lose their license but also be subject to criminal penalties. This would also constrain doctors from administering adequate pain care. The act could increase a physician's legal liabilities again resulting in under-treatment of pain. The bill would give the DEA a mandate to focus on end of life care. The act would empower the DEA to conduct investigations in an area of health care they know very little about. The act would contravene the principles of federalism by overturning the Oregon statue on physician assisted suicide. Finally, the act would reduce innovation and experimentation in the area of palliative care. Physicians would be reluctant to try experimental drug therapies for fear of prosecution (Orentlicher, 2000 pp. 2933-5).

As a result of these criticisms the amended bill died. The lack of knowledge concerning pain care combined with increased government regulation and monitoring, physician fear and political, ideological motives all combined to kill the bill. In its wake the American Pain Society is presently formulating the National Pain Care Policy Act of 2001. This bill would:

> Create a White House Conference on Pain Care. Create a national Center for Pain Research within the National Institute of Health. Provide funding for pain and palliative care education and training through the Agency for Health Care Research and Quality. Create pain care standards within Medicare and Choice plans to ensure payment for appropriate interdisciplinary pain care within these plans. Require an annual report on Medicare expenditures for pain care services, which will require CMS (formerly HCFA) to treat pain medicine as a physician specialty. (Ashburn, 2001 p. 2).

Many of the provisions of this act were incorporated into the National Pain Care Policy Act of 2003 which was submitted to the United States House of Representatives by Representative Rogers (R. Mich.) on April 29, 2003 (United States House of Representatives, 108th Congress 1st Session, 2003). As of mid Fall 2003 no hearings were scheduled. The purpose of the bill was to begin the discussion concerning pain care in the Congress. There is a long legislative road yet to travel. While this budding debate was taking place in the Congress, Bush Administration Attorney General John Ashcroft acted to block Oregon's assisted suicide law by authorizing DEA agents to punish doctors prescribing federally controlled drugs to terminally ill patients (Eggen and Connolly, 2001pp. A1, 3). This was quickly followed by federal judge temporarily blocking the order (Booth, 2001 p. A2) and a United States District Court ruling upholding Oregon's law (Booth, 2002 p. A1). In addition, there were reports in 2003 that federal prosecutors working for the Drug Enforcement Agency were increasing their efforts at prosecuting pain doctors (Kaufman,2003, A1,A5).

One could view these developments as the normal course of political events surrounding public policy issues and to an extent this is true. However, one could also view these events as an example of political ideology driving policy to the detriment of patient pain care management as practiced by physicians across the country. Without an effective patient pain control bill time is wasted and patients suffer. Furthermore the movement toward a patient centered model of health care is impeded. We will draw a few lessons from this experience but first we turn our attention to the stem cell debate.

THE STEM CELL DEBATE

Research on embryonic stem cells represents the holy grail of basic scientific knowledge concerning the future of health care. These cells have the potential to revolutionize the way we look at cures for all sorts of long term illnesses from diabetes to various forms of muscular diseases. The one defining characteristic of these cells is their *pluripotent* nature. They can differentiate themselves into other types of cells. As such they have enormous potential in research affecting areas of the body such as the liver, brain and skin among others. Due to their adaptability they can literally change the course of research on certain diseases. For example, instead of researching better ways to deliver insulin to a diabetic patient in managing the disease, these cells may actually form new pancreatic cells which would replace the old nonfunctional pancreas. This would represent a quantum leap in research for a whole host of diseases and disorders (http://europa.ev.int/comm/research/ quality of life/stemcells/about .html).

There are ethical problems and dilemmas associated with this type of research. Presently, human embryos are the most promising source of stem cells. Usually these can be provided by in vitro fertilization clinics. The process requires the destruction of the embryo to obtain stem cells. This has touched off an ethical and moral debate concerning the wisdom of conducting such research. Most of the controversy is religious in nature. For many organizations there are strong religious objections to destroying these human embryos. The Catholic Church, for example, considers all embryos potential life. In their view, a ban on federal funds for such controversial research will prevent embryo destruction.

Early in the debate, other sources for these stem cells were touted. For example, a report from the United States Department of Health in 2001 indicated that in addition to embryos, stem cells might be obtained from the blood cells of an umbilical cord or from adult tissue such as bone marrow or from mature adult tissue (United States Department of Health, 2000, Report From the Chief Medical Officer's Expert Group…., Executive Summary, 1: Cheng, et al 2003, 131-42). However, new research has indicated that these possibilities were at best optimistic (Weiss, 2003, A6; Chambers et. al, 2003). New research concluded that early research (see, McKay 2000, 361-64) concerning cells derived from human adult bone marrow was misinterpreted. This is important because it may leave embryonic derived stem cells as the only viable source for promising research at this time. This, in turn, implies that the controversy will continue.

The federal government supplies the largest amount of funding for basic research. Privately-funded stem cell research is one approach scientists are pursuing but these resources are limited. Stem cell research was banned by Congress in 1995. The Congress attached the ban to an appropriation bill funding the National Institute for Health. Since 1995 no public funding has been available for this research (AAAS Policy Brief; Stem Cell

Research, September 5, 2003, 2). In 1999, the Department of Health and Human Service indicated that publicly funded research could be conducted on these stem cells as long as the destruction of the embryos for the derivation of these cells were privately funded. They also began to draft regulations for these purposes. In effect the 1999 guidelines permitted research only on cells derived from embryos which were the remnants from fertility treatments at fertility labs. These were donated with consent for research by patients (AAAS Policy Brief; Stem Cell Research, September 5, 2003, 4-6). The United States Department of Health took a position in 2000 that research on nuclear transplantation of human embryo cells should continue subject to safeguards largely modeled after those established in the United Kingdom (United States Department of Health 2000, 44). President Bush's Secretary of Health and Human Services, Tommy Thompson advocated the use of public funds for this research as did some Republican members of the house and Senate, including the influential Senators Hatch from Utah and Thurmond from South Carolina (AAAS Policy Brief: Stem Cell Research, 2003 September 5,). In fact, Senator Hatch distinguished "totipotent" cells which form after fertilization from "pluripotent" cells because the latter could never become a human being. Hatch supports research on "pluirpotent" but not "totipotent" cells (Berger, E 2001). During the 2000 presidential campaign, candidate Bush declared his opposition to any federal funding for this type of research. The stage was set for a dramatic presidential address to the nation in August of 2001. President Bush would make a decision affecting the course of basic research in this nation and all potential future patients who might benefit from this research.

The Bush Decision

President George W. Bush faced a conflicted decision. On the one hand most conservatives in Congress and Christian and evangelical groups opposed any funding for "pluripotent" stem cell research. During his campaign, Bush promised he would not extend public funding to this research. On the other hand, Democrats, liberal groups and a handful of Republican congress people and senators along with Secretary of Health and Human Services, Tommy Thompson and some members of the White House staff urged the president to allow public funding (Berger E.). Faced with a politically and ideologically charged environment Bush attempted to split the issue. He allowed federal funding for research using existing stem cell lines only because the life and death decision had already been made in these instances. Bush prohibited public funding on all other stem cell derivatives existing in fertility clinics nationwide. He also created a new Council on Bioethics chaired by Dr. Leon Kass of the University of Chicago for advice on the ethical and moral implications of this type of research.

The address to the nation is instructive of the president's thinking and influence. The following phrases are taken from the address; "church," "different faiths," "same faith," "potential for life," "a great deal of thought and prayer," "moral hazards," "beginnings of life," "life is a sacred gift from our Creator," "without crossing a fundamental moral line." Bush openly admitted that, "My position on these issues is shaped by deeply held beliefs. I am a strong supporter of science and technology……. I also believe human life is a sacred gift from our Creator.

> (The White House Press Release, 2001). The president went on to encourage stem cell research from placenta and adult human bone marrow sources. The speech demonstrates the faith based influences upon presidential decision making in this area of controversial research and the impact of Christian and evangelical groups upon such decisions in this administration. These groups however were not supportive of the president's limited decision (http://www.americancatholic.org/News/StemCell/default.asp). Of course, how limited the decision was is itself debatable. Kinsley (2003, A25) notes that President Bush was originally expected to ban research involving future embryos but his decision also banned existing embryos in fertility clinics destined for destruction. This, in part, could explain the muted evangelical criticism.

We believe that this decision also reflects Campbell's (2004) definition of ideological entrepreneurship. The president admitted the deeply religious beliefs that influenced his decision-making in his speech. While he did not entirely satisfy his Christian evangelical base he did appease them and offer those interests further opportunity to influence these decisions in the future. Their persistence on this issue makes such an outcome likely.

We note that this ideological policy entrepreneurship is occurring not only in the area of health care policy but also in other areas like basic foreign and environmental policy (Campbell, 2004). Indeed, in environmental policy leading researchers were mobilizing against a Bush administration plan which would block new federal regulations unless the science used to justify these regulations is cleared by a peer review process overseen by the executive branch's Office of Management and Budget. Many scientific researchers believe this would unnecessarily politicize the process (Weiss, 2004, A16). All of this comes in the midst of questions concerning President Bush's general orientation to scientific policy. Both the National Research Council and the Union of Concerned Scientists called upon the Bush administration for greater public access to objective scientific information (Gugliotta and Weiss, 2004, A2). There was also a push for a new source for information on the science of terrorism by the Congress. A two and a quarter million dollar grant by the MacArthur Foundation went to the American Association for the Advancement of Science to link policy-makers who need scientific information with the academic community (Kaufman, 2004, A19).

Regarding the stem cell decision, critics immediately attacked the decision on a number of fronts. First, of the stem cell lines that Bush identified, how many were truly viable for research purposes? Stanford University indicated that of the seventy-eight stem cell lines in the world only about one dozen might be in good enough shape for research purposes (PBS Online Newshour Update, December 11, 2002). Second, are these cells genetically diverse enough for research purposes? Third, what is the safety of the existing cells concerning future human implantation? Fourth, the viability of the long run research in this area might be significantly slowed by utilizing only these existing stem cell lines. In fact, The National Academy of Science in September of 2001 concluded that in the long run new embryonic stem cell lines will be necessary (AAAS Policy Brief: Stem Cell Research, 2003). Fifth, many critics predicted that the United States would lose basic research scientists to other nations that have more liberal laws regarding stem cell research. This has not happened as of 2003 largely because the states have stepped in to create research friendly zones for this particular type of research. California has enacted legislation to attract these researchers (Gershon, 2003). Stanford University received a twelve million dollar grant from an anonymous donor to establish an institute that will create new stem cell lines (PBS Online Newshour Update, 2002, December 11). The effect of the Bush ban upon scientists leaving the nation to conduct research elsewhere is largely determinative of how long the ban lasts and the research questions current experiments lead to. Clearly, the jury is still out.

There were other problems with the ban that apparently were not considered in the decision making process. Fins and Schacter (2002: 269-70) argue that the decision is unwise for two reasons. The first involves patent law and the second principles of informed consent. Regarding patent law the authors note:

> ....Ordinarily, upon the expiration of a patent, the patentee may no longer effectively prohibit others from formulating the product. Patent law temporarily limits the monopoly so that the invention reaches the public. The president's policy subverts the construct of patent law because it prohibits federally funded researchers from using the disclosed knowledge to create new cell lines even after the patents expire, as such researchers will not have critical access to the ingredient they need to exploit the knowledge. This will be to the detriment of ensuing research and potential utilization in therapeutic product development.

Fins and Schacter further note that the Bush policy will provide an advantage to the initial holders of these patents secured before August 2001. The artificial market constraint imposed by the Bush decision will work against the purpose of our anti trust and patent laws to promote innovation and enhance consumer welfare. Thus the decision will ultimately affect future patients in this area. As Fins and Schacter (271) argue, "To restrict scientific enquiry through restraint of markets upon which biomedical science increasingly depends could deprive a generation of patients of clinical advances..."

The patent argument is becoming central to these debates. In November of 2003 Senator Sam Brownback (R-Kan.), a pro life supporter, added language to a Senate Appropriations bill which would ban patents on human embryos. Brownback followed the lead of Rep. David Joseph Weldon (R-Fla) who, in a bill passed by voice vote in the House, banned patents on human organisms (Weiss, 2003, A4; Weiss, 2003, A19). The Weldon bill has been stalled in the Senate largely due to the efforts of Senator Hatch (R-Utah) (Novak, 2003, A31). These actions touched off a philosophical debate on the role of free markets and nascent human development. It marks the first time Congress has intervened in such issues and potentially pits the Congress against the U.S. Courts.

A second problem involves the principle of informed consent. Are the consent forms which existing donors signed adequate after the Bush decision? Many of these donors were assured that their identity would be lost in a sea of research consent forms. However, with the Bush limitation these donors will be far more concentrated and perhaps conspicuous. Did these donors fully appreciate the economic value of their donations? The authors (273) ask, "What would happen if individuals who had consented to these procedures reconsidered their decision based on the high-profile nature of these sanctioned cell lines and the public persona of their DNA?" How might these circumstances affect future donors in this area? Might it chill basic research? If these questions are answered in the positive they will undoubtedly affect future patients by slowing research.

There are other problems concerning the President's Council on Bioethics. The Council consists of eighteen members but only three scientists and two clinicians. There were charges that the Council has strong conservative, theological and religious leanings and might be skeptical of science and technology in general (Check, 2002, January 24, pp.351-352). The president's panel recently called for a ban on radical reproductive procedures such as creating animal/human hybrids. However, draft reports initially included politically divisive language indicative of a heavy philosophical and scientific debate among the members (Weiss, 2004, A2). Given the nature of the Bush address heavily laden with religious belief and conviction this should not be a surprise. In fact the Bush Administration has come under attack for its misuse of science in a number of policy areas (Weiss, 2003). In addition, the president has recently created new concerns about his objectivity with respect to news in general. The president noted in an interview with FOX news, "The best way to get the news is from objective sources. And the most objective sources I have are people on my staff who tell me what's happening in the world" (Kinsley, 2003).

Objective information is extremely important in this basic area of research. One of the dangers already occurring is the pollution of objective, peer reviewed studies with other reports, news articles and studies of dubious distinction. For example in 2002 the United States Senate Judiciary Committee held hearings on cloning. During this hearing senators and the news media routinely utilized articles and reports of dubious scientific validity. Many of these articles and reports had not been subject to scientific peer review procedures. In fact Stanford biologist Irving Weissman warned the senators that such reports should not be used because they might unduly influence the legislative debate (Check, 2002, February 14 p.722). This is a major problem in a democratic society which relishes first amendment principles. The danger of un-objective reports crowding out scientific reports is ever present and growing. The problem increased when the president joined the debate against therapeutic cloning and argued that, "Research cloning would contradict the most fundamental principle of medical ethics – that no human life should be exploited or extinguished for the benefit of another" (Check, 2002, April 18). If such an information environment is allowed to continue belief will trump science to the detriment of patient care. Irving L. Weissman noted in the *New England Journal of Medicine* (2002, 1578):

> Unfortunately, there are few in Congress or the President's council who can evaluate the scientific and medical issues in order to make an appropriately informed decision. Too often in recent Senate hearings, the views expressed by senators have been based on articles in newspapers and popular magazines rather than reports of the National Academies or articles in peer-reviewed journals. Some journalists are failing the public trust by publicizing findings that have not been published in the scientific literature or independently replicated.

How do these developments in the basic stem cell research area relate back to the newly emerging patient centered health care model? First, scientific technology and inquiry has reduced the amount of time between basic research discoveries and their clinical applications.

Basic scientific discoveries are emerging more rapidly today than twenty-five years ago. The human genome project is indicative of such a pace. When non-scientific obstructions are placed upon basic scientific research, the time to discovery and clinical application will be interrupted. Needless suffering will occur because of these obstructions. This sets back the newly emerging paradigm for it reduces the opportunities for patient control and empowerment with respect to a whole host of diseases and disorders. How can patients be empowered in a society that places impediments on basic scientific research, artificially alters the information environment and interferes with the free market results of basic research? Such a society is more in keeping with the old paternalistic health care model rather than the new patient centered one. What are the lessons from the ideological based decisions in the area of pain care management and basic stem cell research?

Conclusions and Recommendations

It would be easy to state that ideological or religious based decision making should be tempered by objective scientific inquiry. Wishing will simply not bring this about. There are some basic suggestions that might help further the cause of objectivity in these most important areas. In the area of pain care the importance of education cannot be overstated. Health care professionals and their associates including insurers, state medical boards, law enforcement agencies and general practice physicians must be better educated in pain control especially as it affects life and death issues.

As Dr. Stephen Long noted with respect to pain care:

> If you compare other medical specialties, we get the patients with problems that many primary care doctors do not often see or poorly understand, or haven't been educated to treat properly. Not because they are ill informed but because they do not have any specialty training in pain management. Traditionally, little if any education regarding pain is presented to medical students or while in residency. They rely on what they read or hear historically. So pain specialists become the first line of defense for many of these patients after they have gone from doctor to doctor. We get a lot of referrals from orthopedic and neurosurgeons who are more specific and more geared toward an accurate pain diagnosis. They are asking us (pain care specialists) to intervene and to implement some changes and take care of those patients. As pain care specialists we make many of the initial diagnoses and provide education for not only the patient but the health care provider who referred the patient. We also direct the physical therapist who may not be accustomed to treating a specific condition in a chronic pain patient. We help the pharmacist because the pharmacist may or may not be used to prescribing the doses of medication for the chronic pain indications. There is also a lot of time spent educating the insurance provider as well. The pain care specialist spends an enormous amount of time providing education and that time is not reimbursable.

If Dr. Long is correct about the educational role of the pain care specialist how can we as a society increase the education concerning pain care to empower the patient? First, medical schools can and should make pain care a primary field of study. This will increase the education of doctors who later serve as advisers to drug enforcement organizations, state medical boards and insurance providers. Second, pain control standards should be nationally known and publicized. The JCAHO standards for pain management are beginning to set standards within institutions for this purpose. Third, congressional hearings should be held on the newly proposed National Pain Care Policy Act separate from euthanasia or 'Right to Life' agendas. Fourth, yearly workshops and symposiums on standards and new developments in pain care should be provided to all health care providers. Fifth, the best scientific evidence available should be the foundation for information when legislating public health issues thus avoiding public policy decisions based on alternative or ideological agendas. A process for dissemination of public information based on the scientific evidence can become a citizen's expectation rather than the exception. Throughout the 1990's the federal government, through the Agency for Healthcare Research and Quality (AHRQ) has provided research support and policy guidance, in collaboration with experts from the public and private sectors. Despite a federal (The Healthcare Research and Quality Act of 1999, Part B, TitleIX, Section 911(a)) mandate that AHRQ identify systems to assess research results, rating the strength of the scientific evidence underlying health care practice, and make systems widely available, this model has been largely ignored in pain care legislation and stem cell policies. (www.infopoems.com/)

What specific suggestions can be gleaned from the stem cell decision? Any congressional or presidential advisory body should have a *balanced* membership between scientific researchers in the area of inquiry and a mix of ethicists, physicians and journalists otherwise such a panel should be suspect. The staff of such an advisory board or panel should employ legal help in a variety of areas including privacy and economic concerns. Information at any congressional hearing or nationwide symposia should be categorized by order of its scientific value. Over the last 15 years the medical community has used a systematic method for evaluating several independent clinical trials and outcomes. This system allows for more objective appraisal of evidence, which may lead to resolution of some present disagreements. In addition, it can reduce the probability of false negative results and help speed up the process to the introduction of effective treatments (Egger and Smith 1997). The science should be presented based on systematic research-based evidence rather than on expert opinions or clinical experience thereby minimizing bias and ideology. A hierarchy should be established with scientifically, peer reviewed research receiving the highest level or category of information. Participants, journalists, members of congress and the general public should be made aware of the levels of information categorization.

These suggestions, while basic in nature, can go a long way in helping to temper the impact of ideology upon public policy decision making. We can never eliminate ideology nor should we. However, in the twenty-first century the pace of scientific discovery and medical advances should not be impeded by fear and misinformation. Ideology must be balanced by scientific knowledge to satisfy the emerging patient centered paradigm in health care. The future of patient care demands no less.

Acknowledgement

Paper prepared for The Annual Southern Political Science Association Conference, New Orleans, January 7-10 2004. The Authors would like to thank Robert Gardner for his assistance in the research for this article.

References

American Association for the Advancement of Science (AAAS). Online. (2003, September 5). AAAS Policy brief: Stem cell research. Center for Science, Technology, and Congress *http://www.aaas.org/spp/cstc/briefs/stemcells/*

Ashburn, M. A. 2001. "National Pain Care Policy Act of 2001." *American Pain Society Bulletin. 11*: 2.

Ashburn, M. A. 1996. "Effort to Establish a National Pain Research Center Progresses." *American Pain Society Bulletin. 6*: 8-9.

Ashburn, M. A. and P.S. Staats. 1999. "Management of Chronic Pain." *Lancet 353*: 1865-9.

Berger, E. (2001, July 11). Research avenue adds fuel to stem cell controversy. *Cnn.com http://cnn.com/2001/HEALTH/07/11/stem.cell.fact*

Booth, W. 2002. "Oregon's Assisted Suicides Upheld." *The Washington Post. 18* April. A1.

Booth, W. 2001. "Judge Blocks Sanctions Over Assisted Suicide." *The Washington Post.* 9 Nov. A2.

Campbell, C. (2004) Unrestrained ideological entrepreneurship in the Bush II advisory system: An examination of the response to 9/11 and the decision to seek regime change in Iraq. In C. Campbell and B.A. Rockman (Eds.). *The George W. Bush Presidency: Appraisals and Prospects*. (pp74-75). Washington, DC: CQ Press.

Carr, D. B. (1999). Pain in the Media. (Address delivered at *Pain In The Public Eye: 18th Annual Scientific Meeting of the American Pain Society.* Fort Lauderdale, Florida, October 22, 1999.

Chambers, I., Colby, D., Robertson, M., Nichols, J., Lee, S., Tweedie, S., Smith, A. (2003, May 30). Functional Expression Cloning of Nanog, a pluripotency Sustaining Factor in Embryonic Stem Cells. *Cell.*Vol. *113*: 643-655.

Check, E. (2002, January 31). Senators square up over cloning. (News) *Nature.* Volume *415*, p. 461.

Check, E. (2002, January 24). Panels' conflicting views cloud legal future of human cloning. (News) *Nature.* Volume *415*, pp. 351-352.

Check, E. (2002, February 14). Cloning agenda 'skewed' by media frenzy. (News) *Nature.* Volume *415*, p. 722.

Check, E. (2002, March 07). Call for cloning ban splits UN. (News) *Nature.* Volume *416,* p. 3.

Check, E. (2002, April, 18). Bush rallies opposition to cloning. (News) *Nature.* Volume *416*, p. 669.

Cheng, L., Hammond, H., Ye, Z., Zhan, X. and Gautam, D. (2003). Humanadult marrow cells support prolonged expansion of human embryonic stem cell culture. *Stem Cells 21,* Volume *2*: 134.

CNN.com. (2001,August 10, 2001). Bush to allow limited stem cell funding. (*http://cnn.com/2001/ALLPOLITICS/08/09/stem.cell.bush*)

Cohen, R. (2001). An Ethical Travesty. *The Washington Post.* 14 August. Op-Ed.

Connolly, C., Gillis, J. and Weiss, R. 2001. "Viability of Stem Cell Plan Doubted." *The Washington Post. 20* August. A1, 4.

Dennis, C. (2002, March, 07). Stem-cell reverse angers Australian biologists. (News) *Nature.* Volume *416*, p. 4.

Department of Health, Office of the Chief Medical Officer. (2000, June). A report from the Chief Medical Officer's expert group reviewing the potential of developments in stem cell research and cell nuclear replacement to benefit human health.

Eggen, D. and Connolly, C. (2001). Ashcroft Ruling Blocks Ore. Assisted Suicide Law. " *The Washington Post. 7* Nov. A1, 3.

Egger, M. and Smith, G. D. (1997). Meta-analysis: Potential and promise. *BMJ.* No. 7119, vol*315*.

Europa European Union Online. (2000) About stem cells – Seven important questions on stem cells. (*http://europa.ev.int/comm/research/qualitylife/stemcells/about.htm/*)

Finns, J.J and Schacher, M. (2002) Patently Controversial: Markets, morals, and the president's proposal for embryonic stem cell research. *Kennedy Institute of Ethics Journal.* Vol. *12*, 3,:265-278.

Goldstein, A. (2001, October 1). Terrorism tied to jump in pain problems. *The Washington Post.* Section A, pp. 1and 6, Col 1.

Gershon, D. (2003, April 24). Stem-cell research. *Naturejobs 422,* pp. 928-929.

Gugliotta, G. and Weiss R. (2004, February 19). President's scientific policy questioned. *Washington Post,* Section A p2.

Hansen, G. (2000). Assessment and Management of Pain: Letters. *The Journal of the American Medical Association. 284*: 2317-2318.

http://www.americancatholic.org/news/StemCell/default.asp

Hyman, C. S. (1996). Pain management and disciplinary action: How medical boards can remove barriers to effective treatment. *Journal of Law, Medicine and Ethics. 24*: 338-43.

InfoPOEM, Inc. *http://www.InfoPOEMS.com*

Joint Commission on Accreditation of Healthcare Organization. August, 1999. Pain assessment and management standards. [Online]. *http://www.jcaho.org/standards/pm*

Kaufman, M. (2003, Dec. 12). Worried Pain Doctors Decry Prosecutions. *Washington Post.* Section A p.1,4.

Kaufman, M. (2004, February 9). Linking Lawmakers, Scientific Knowledge. *Washington Post,* Section A, p.19.

Kinsley, M. (2003, October 17). Bush's Filtered News. *Washington Post,* Section A, p. 29.

Kinsley, M. (2003, November 31). Kabuki and Stem Cells. *Washington Post,* Section A, p25.

Knight, J. (2002, February 14), American Red Cross turns it's back on stem-cell grant. (News). *Nature*. Volume *415*: pp. 719.

Long, Stephen. (Director of Pain Management Center) (2001, October 16) *Steve Long Interview* Richmond, Virginia.

Max, M. B. (1990) Improving Outcomes of Analgesic Treatment: Is Education Enough?" *Annals of Internal Medicine, 113*. 885-888.

McKay, R. (2000, July 27). Stem cells – hype or hope. *Nature, 406*, pp.361-364

Morris, D. B. (1991). *The Culture of Pain*. Berkeley: University of California Press.

Novack, R. D. (2003, December 15). New Jersey's End Run On Cloning. *Washington Post*. A31.

PBS Online NewsHour Update. (2002, December 11, 4:45pm EST) *Stanford University to create new stem cell lines.*

Orentlicher, D. (2000). Legislation and End of Life Care. *Journal of the American Medical Association. 283*:2933-5.

Orentlicher, D., Caplin, A. (2000). The Pain Relief Promotion Act of 1999: A serious threat to palliative care. *Journal of the American Medical Association. 283*:255-258.

Philipkoski K. (2001, Aug. 9, 1:40 pm). Bush OK's stem cell funding. *Wired News*. (*http://search.wired.com/wnews/default.asp?query=Bush%20OKs%20Stem%20Cell%20 Funding*)

Post, L. F., Blustein, J., Gordon, E. and Dubler, N. (1996). Pain: Ethics, culture, and informed consent to relief. *Journal of Law, Medicine and Ethics. 24*: 348-59.

Reuters Health. (2000, Jan.17). Medical ethicists claim Pain Relief Promotion Act of 1999 is bad nedicine. *Medscape.com http://primarycare.medscape.com/reuters/2000*

Rowlingson. J. (Professor of Anesthesiology and Director of Pain Management Services at the University of Virginia), (2001, March 14) *John Rowlingson Interview*. University of Virginia.

Weissman, I.L. (2002). Stem Cells – Scientific, medical, and political issues. *New England Journal of Medicine, 20,* Vol. *346*: 1576-1579.

Weiss, R. (2003, August 8). Bush misuses science, report says: Democrats say data are distorted to boost conservative policies. *Washington Post,* Section A, p. 15.

Weiss, R. (2003, October 13). Stem cell research is questioned: Bone marrow's potential overstated, study shows. *Washington Post,* Section A, p. 6.

Weiss, R. (2003, November 17). Funding Bill gets Clause on Embryo Patents: ban Inserted into appropriations measure Stirs Political, Philosophical Debate. *Washington Post.* A4.

Weiss, R. (2003, November 25). Hill Negotiators Agree to Bar Patents for Human Organisms. *Washington Post.* A19.

Weiss R. (2004, January 15). Peer review plan draws Criticism. *Washington Post,* Section A p19.

Weiss, R. (2004, January 16). Bioethics Panel Calls for Ban on Radical Reproductive Procedures. *Washington Post,* Section A p. 2.

The White House, Release from the Office of the Press Secretary. (2001, August 9, 8:01pm). Remarks by the President on stem-cell research. *http://whitehouse.gov/news/releases /2001/08.*

In: Frontiers in Stem Cell Research
Editor: Julia M. Spanning pp. 135-151 ISBN 1-60021-294-8

Chapter V

MULTIPOTENT STEM CELLS FROM SKIN DERMIS: BASIC BIOLOGY AND THERAPEUTIC IMPLICATIONS

Chunmeng Shi**, Ying Zhu, Yongping Su, Chengji Luo and Tianmin Cheng
Institute of Combined Injury, State Key Laboratory of Trauma, Burns and Combined Injury, College of Preventive Medicine, Third Military Medical University, Chongqing, China

ABSTRACT

Skin stem cell biology has been a focus of increasing interest in life science. Given its easy accessibility, the broad developmental potential of stem cell populations from skin has provided more wide practical implications. Skin is the biggest organ in the body and is constantly renewed. This process of renewal necessitates the presence of adult stem cells. To date, research has focused on the application of epidermal stem cells. Recent research also shows there are at least three different types of other stem cells (mesenchymal, neural and hematopoietic stem cells) that have been identified from the dermis and hair follicles. Mesenchymal stem cells from bone marrow are the best studied adult multipotent stem cells and have shown promising therapeutic implications in regenerative medicine. The multipotent stem cells from the dermis have the potential to produce distinct cell lineages, such as fibroblasts, follicle dermal cells, endothelial cells, marrow stromal cells, osteocytes, chondrocytes, muscle cells, and even neural cells. Recently research indicates that the dermis may represent a novel source of mesenchym stem cells. In this chapter, recent findings regarding the biological properties and potential clinical use of dermis-derived multipotent stem cells with mesenchymal

* Correspondence concerning this article should be addressed to Dr. Chunmeng Shi, Institute of Combined Injury, College of Preventive Medicine, Third Military Medical University, 30 Gaotanyan Street, Shapingba District, Chongqing, 400038, China

potential are summarized. Recent advances of the neural stem cells and hematopoietic stem cells from the dermis are also discussed.

Keywords: Adult Stem Cells, Skin, Trauma, Hematopoiesis, Malignant Transformation

INTRODUCTION

Adult multipotent stem cells can be isolated from various tissues and can be used to circumvent immunological problems. These advantages make them an attractive source for cell-based therapies [1-3]. The skin is the biggest organ in the body and is critical to protect the body against environmental hazards. Adult skin consists of epidermis, dermis and appendages such as hairs and glands that are linked to the epidermis but project deep into the dermal layer. The skin is constantly renewed and remodelled with a strict balance between proliferation and differentiation [4,5]. In humans, the epithelial part of the epidermis is completely renewed once every three weeks. Similarly, the dermis is constantly remodelled with new fibres being produced, while old ones are degraded. Hence, renewal of the epidermis and dermis necessitates the presence of adult stem cells. In the traditional views, skin stem cells, usually preferred to epidermal stem cells, which are unipotent, meaning that they generate a single lineage. Actually, there are more stem cells in the skin than the epidermal stem cells.

Mesenchymal stem cells with broader developmental potential are firstly identified from bone marrow and have shown promising future in regenerative medicine [6,7]. In addition to marrow, other sources of stem cells with mesenchymal potential are still required. There are some evidence that mesenchymal stem cells can be isolated from periosteum, trabecular bone, synovium, skeletal muscle, lung and deciduous teeth [8-15]. However, these sources have some limitations for further usage. The bottom 90% of skin known as the dermis contains several types of mesenchymal cells among which are fibroblasts. Recently, multipotent stem cells with mesenchymal potential have been isolated and identified from the dermis and adipose tissues in the skin. These stem cells have been shown to have similar differentiation potential with cells from bone marrow in terms of yield, growth kinetics, cell senescence, multi-lineage differentiation capacity, and gene transduction efficiency. Skin stem cell biology has been a focus of increasing interest in life science [16,17]. Given its easy accessibility and high developmental potential, multipotent stem cells in the skin will provide an alternative source of adult mesenchymal stem cells and have clinical importance and practical implications for several areas including skin reconstitution and tissue engineering. In this chapter, the biological properties and clinical implications of dermis-derived multipotent stem cells, referred to as dermal multipotent cells (DMCs) are summarized and discussed.

CONCLUSION

1. Isolation and Characterization of Adult DMCs

Morphology

Multipotent stem cells can be isolated from dermis by their adherence to tissue culture surfaces. In general, enzymatically dissociated dermal cells are generally cultured in basal medium such as Iscove's Modified Dulbecco's Medium (IMDM) in the presence of 10% fetal bovine serum (FBS). Primary cultures are usually maintained for less than 24 hours and only a very small percentage of the cells adhere to the tissue culture substrate. Some single cells form colonies with one layer and a few single cells form colonies with more than two layers. These colonies confirmed by multilineage differentiation are regarded as plastic-adherent DMCs. By clonal analysis, there are at least two cell subpopulations in different size and morphology in single sorted DMCs like that in bone marrow mesenchymal stem cells: most of large spindle-shaped cells or flat cells and a small percentage of small round cells (less than 5%). There are some differences between the two cell populationss both in growth kinetics and differentiation potential. The doubling time of small round cells is 24h, whereas of the large cells is about 48h. When cultured respectively, the small round cells grow faster and produce more colonies than the large cells. The small round cells gradually change their morphology into spindle-shaped or flat cells while the large cells can't change into small round cells in culture. The small cells in the filtrate are about 7 μm in diameter and have a high nucleus-to-cytoplasm ratio than large cells. In contrast, large cells retain on the filter are about 35 μm in diameter and contain a large number of unidentified vacuoles. Analysis of differentiation potential also shows that the small round cells differentiate more extensively into osteocytes and chondrocytes than the large cells in osteogenic and chondrogenic medium. These results suggest that the small round cells are more primitive than large cells [18-20].

Surface Antigens

Considerable effort has been expended on the identification of specific surface markers for selection, detection and testing of DMCs. Several monoclonal antibodies have been used for the characterization and isolation of DMCs. By flow cytometric analysis, rat DMCs are negative for lineage-specfic surface markers of including pan-cytokeratin, cytokeratin19, Factor VIII, CD31, CD34, α-smooth muscle actin (α-SMA) ,desmin, type II collagen and NF-200, but are positive for CD90, a marker for thymocytes and peripheral T lymphocytes and are also positive for some adhesion molecules, like CD59, CD44, intercellular adhesion molecule-1 (ICAM-1) and vascular cell adhesion molecule-1 (VCAM-1). Mesenchymal stem cells from human dermis are also positive for CD90 and are negative staining for CD3, CD4, CD8, CD11c, CD33, CD36, CD38, CD45, CD117, Glycophorin-A, and HLA DR-II [20]. Much valuable information can also be gained from a systematic analysis of cell surface molecules on DMCs.

Table 1. The gene lists associated with diverse cellular lineages in DMCs by DNA microarray (partial)

Cell lineage	Representative examples of associated genes
A: Osteoblast	osteonectin, osteopontin, chondroitin sulfate proteoglycan 2 (versican), biglycan, bamacan, collagen type I
B: Muscle	smooth muscle-specific 17beta-hydroxysteroid dehydrogenase type 3; skeletal muscle myosin light chain kinase; skeletal muscle voltage-sensitive sodium channel alpha subunit; muscle type calpain p94; skeletal muscle myosin heavy chain; striated muscle alpha-tropomyosin; smooth muscle cell LIM protein; smooth muscle myosin; skeletal muscle elongation factor-2 kinase; phosphofructokinase muscle isozyme; CELF, SM22
C: Fibroblast	Prolyl 4-hydroxylase, fibronectin, basic fibroblast growth factor, keratinocyte growth factor, fibroblast growth factor receptor, vimentin, collagen type I, III
D: Epithelial cell	epithelial membrane protein 1, epithelial cell transmembrane protein antigen precursor, epidermal growth factor precursor, heparin-binding EGF-like growth factor , mucin, parathymosin and thymosin 4 and 10
E: Endothelial cell	VEGF, VCAM-1, syndecan , microvascular endothelial differentiation gene 2, endothelial receptor for oxidized low-density lipoprotein
F: Neural cell	neural visinin-like Ca2+-binding protein type 3; neuron-specific enolase; neuronal nitric oxide synthase; GABA receptor-associated proteins, NCAM, brain-derived neurotrophic factor (BDNF), syntaxin 7, glial fibrillary acidic protein (GFAP), adrenomedullin, myelin protein SR13 (growth-arrest-specific Gas-3 homolog), Neurotrimin; secretogranin II; synapse-associated protein 102; latexin; kexin-like protease PC7A; NMDA receptor
G: Pancreatic cell	pancreatic eukaryotic initiation factor 2 alpha-subunit kinase (PEK), pancreatic secretory trypsin inhibitor-like protein (PSTI), pancreatic islet cDNA Rattus norvegicus cDNA similar to rapamycin-binding protein FKBp-13, insulin II gene, IRS-1 mRNA for insulin-receptor, mRNA for glucose-dependent insulinotropic polypeptide
H: Hepatocyte	liver glycogen phosphorylase enzyme; liver IL-6 receptor ligand binding chain; liver specifc transcription factor LF-B; liver nuclear protein P47; liver a-L-Fucosidase
I: Hematopoietic cell	hematopoietic lineage switch 2 related protein (Hls2-rp), hematopoietic prostaglandin D synthase, bone marrow stromal cell antigen 1 (BST-1), T cell receptor eta chain, CD14, CD24,IL-3, IL-6

Gene Expression Profiles

In order to obtain a better understanding of DMCs, the gene expression profile is assessed by DNA microarray analysis and the harbinger of a potential cell type is viewed by simultaneous expression of a panel of lineage-related genes. The result shows cultured DMCs simultaneously express genes associated with diverse mesenchymal cell lineages (Table 1.), including osteoblast, muscle, and fibroblast, suggesting the existence of a pluridifferentiated mesenchymal progenitor cell type. Cultured DMCs also simultaneously express transcripts for epithelial cell, endothelial cell, neural cell, pancreatic cell types and hepatocyte as well, widening its transcriptomic repertoire. Furthermore, DMCs express transcripts of hematopoietic cells. Moreover, DMCs also express transcripts of a series of cytokines and extracellular matrix molecules (Table 2.), including vascular endothelial growth factor (VEGF), platelet derived growth factor (PDGF), hepatocyte growth factor (HGF), transforming growth factor-beta (TGF-β), ICAM-1, VCAM-1 and Fibronectin [21,22].

Table 2. The gene lists associated with cytokines, receptors and surface molecules in DMCs by DNA microarray (partial)

Gene Classification	Representative examples of associated genes
Cytokines	IL-15; IL-6; IL-3beta; basic fibroblast growth factor (bFGF); transforming growth factor alpha (TGF-alpha); TGF-beta 1; TGF-beta3; Vascular endothelium growth factor (VEGF); Insulin like growth factor 1 (IGF-1); latent TGF-beta binding protein2 and 3; NGF inducued factor A; High mobility group protein (HMGP); hepatocyte growth factor (HGF); platelet-derived growth factor (PDGF) A and B; heparin-binding EGF-like growth factor; epidermal growth factor precursor
Receptors	Type III TGF-beta receptor; TGF-beta type II receptor; TNF receptor; transferring receptor; IL-1 receptor related protein; hepatocyte growth factor receptor; IL-1 receptor type 1 and 2; IL-3 receptor; IGF-I receptor; bFGF receptor; c-met; PDGF receptor alpha and beta
Extracellular matrix and adhesion molecules	collagen type I; collagen type III; fibronectin; integrin; syndecan; ICAM-1; VCAM-1; praline-rich proteoglycan; chondroitin sulfate proteoglycan 2; CD44; procadherin2,4,5 and 6

Growth Kinetics

When cultured in vitro, there is no obvious difference of growth patterns between cells of the third and tenth passage. Following culture of the third and tenth passage, an initial time of 4~6 days is observed during the log growth phase, when cells are successively cultured for 20 passages, the initial time became 8 to 10 days. Under the in vitro conditions, DMCs retained the osteogenic, chondrogenic, and adopogenic differentiation potential after extended culture of more than 20 passages. In order to compare the osteogenic potential of dermal multipotent cells at different passage, osteogenic differentiation potential of the third, tenth and twentieth passage are compared by specfic stains of alkaline phosphatase activity (AP activity) and calcium composition. After induction, stains for alkaline phosphatase activity and calcification of the third, tenth and twentieth passage do not change obviously [23-25].

Differentiation

The differentiation of multipotent stem cells into bone, cartilage and fat has been described and characterized by multiple laboratories. Following induction, DMCs show multilineage differentiation capacity. When cultured in osteogenic medium in the presence of BMP-2, dexamethasone and fetal bovine serum, the alkaline phosphatase activity is increased at day 14 and about 55% DMCs produce mineral associated with the matrix and are stained with alizarin red on day 21. When cultured in chondrogenic medium in the presence of TGF-β, dexamethasone and fetal bovine serum for 21 days, about 25% cells are positive for collagen II and alcian blue staining which indicate that some DMCs are changed into chondrocyte phenotype. When cultured in adipogenic medium in the presence of isobutyl-methylxanthine (IMBX), dexamethasone and fetal bovine serum for 21 days, about 65% DMCs produce Oil red-positive cells. Dermal multipotent cells also produce nestin and NF-200 in neural medium in the presence of rentionic acid after treatment for 21 days. The neural differentiation levels in DMCs averages 10%. We have also shown that the morphology of DMCs can be changed into epithelial cells-like cells when cultured in medium containing ITS (Insulin-transferrin-selenium). Immunohistochemistrial staining shows that a very low percentage of cells (less than 10%) are positive for insulin. This result further indicates that DMCs have broader differentiation capacity [19,22].

Immunological Properties

The immunological properties are important for the successful engraftment of allogenic grafts. Bone marrow-derived MSCs have been proved to exhibit immunosuppressive effects on lymphocytes. In our study, DMCs are found to express mRNA for MHC class I and MHC class II molecules. The in vivo studies shows that There is no significant difference of the percentages of CD2 (expressed on the whole lymphocytes) and CD25 (activated lymphocytes) positive lymphocytes in peripheral blood at 7, 14, and 21 days after systemic transplantation of DMCs in rats. The IL-2 level in peripheral serum also is not significantly changed. By histological observation, no obvious graft versus host disease (GvHD) is observed in the skin, liver and intestine tissues after transplantation. These results indicate that DMCs have low immune rejection activities and don't induce an obvious activation of allogenic lymphocytes [21].

2. Origin and Location of Multipotent Stem Cells in Dermis

For studies on skin development, regeneration and repair, the location of stem cells is a very important thing. Stem cells in epidermis have been widely studied. In normal epidermis, stem cells are usually found in the epidermal basal layer in clusters [26,27]. Scattered epidermal stem cells can also be seen in the spinous and granular layers between the basal layer and the stratum corneum [28]. Recently, it is identified that the upper region of the outer root sheath of mouse hair follicle contains multipotent cells that respond to morphogentic singnals to generate multiple hair follicles, sebaceous glands, and epidermis, i.e., all lineages of the hairy skin [29]. Taylor et al. further demonstrated that follicle epithelial stem cells located in the upper follicle bulge [30]. By these data, the follicle outer root sheath has already been tested as a specific source of epidermal stem cells. Alternatively, it is possible that skin may be widely "seeded" with migrating follicle-derived progenitors, particular in young skin [31]. There are close developmental and anatomical parallels between epithelial tissue and dermal tissue [32]. The epidermis and dermis are both formed of multiple layers with distinct cell types from embryogenesis. Anatomical analysis of the interactions between the hair follicle and surrounding skin have showed that there are parallel relationships between the out root sheet of hair follicle which merges with the epidermis and dermal sheath tissue which connects with the papillary layer of the dermis. Although the origin of dermal multipotent cells has not been well assessed, hair follicle dermal cells are considered as embryonic type cells retained in an adult system and may be a source of dermal stem cells. When the hair follicle base is experimentally amputated, residual cells of the dermal sheath replace the main inductive dermal element the dermal papilla, and restore hair growth in a unique example of mammalian regeneration. With a parallel comparative study of epidermal cells and dermal cells, dermal sheath cells can become wound healing fibroblasts after injury [32]. It is hypothesized that hair follicle is a nich for stem cell populations. There is increasing evidence that hair follicle dermal cells are multipotent and have the potential to differentiate into produce ostecytes and adipocytes. This support that follicle dermal sheath may represents a strictly origin of these multipotent cells. However, sporadic multipotent cells may also exist in non-follicle dermis. Evidence also shows that cultured multipotent adult stem cells may represent endogenous multipotent precursor cells. Skin derived progenitors (SKPs), cultured adult stem cells from the dermis that generate both neural and mesodermal progeny, exhibit properties similar to embryonic neural-crest stem cells. These endogenous SKPs can first be isolated from skin during embryogenesis and that they persist into adulthood, with a niche in the papillae of hair follicles. Lineage analysis further indicates that follicle dermal papillae contain neural-crest-derived cells. Thus, adult multipotent stem cells may represent an endogenous embryonic precursor cell that arises in peripheral tissues such as skin during development and maintains multipotency into adulthood [33,34].

3. Plasticity of DMCs in Hematopoiesis

Hematopoietic stem cell (HSC) transplantation is used increasingly for treatment of hematologic and solid malignancies. Mesenchymal stem cells (MSCs) from marrow have the potential to develop into marrow stroma. Successful engraftment, survival and proliferation of HSCs depend on interactions with their microenvironment provided by bone marrow stromal tissue. Cotranplantation of MSCs and HSCs have obtained a fast recovery of hematopoiesis [35-37]. However, auto-transplantation of marrow MSCs are limited especially when the bone marrow has been injuried in some kinds of hematopoitic diseases characterized by stromal failure, such as multiple myeloma, spontaneous cytopenias and aplastic anemia. Under these conditions, there required other sources of multipotent stem cells for replacement therapy instead of marrow MSCs. Since DMCs shared similar properties with MSCs, the functional analysis of DMCs on hematopoiesis is performed in experimental studies. Under the conditions without granulocyte/macrophage colony stimulating factor (GM-CSF) supplement in semisolid medium, DMC supportive layer supports the formation of colony forming unit of granulocyte/macrophage progenitors (CFU-GM) and colony forming unit of erythroid progenitors (CFU-E) colonies. Under the conditions with GM-CSF supplement in medium, the counts of CFU-GM and CFU-E colonies on DMC supportive layer significantly increase than control without DMC layer. DMCs have been found to produce several growth factors and/or adhesion receptors for hematopoietic cell development. These results show that DMCs support and promote the growth of CFU-GM and CFU-E in vitro. In order to observe the effects of DMCs on hematopoiesis in vivo, a model of hematopoisis damage by exposing rats to sublethally 5Gyγγand lethally 8.5Gy γ-ray-irradiation is used respectively. DMCs are infused into the tail veins of rats after irradiation. For animals received 5Gy irradiation, the white blood cell (WBC) count at 7 days reduces significantly than those of the normal animals due to the hematopoitic damage caused by irradiation. In DMC transplantation group, the WBC count at 15 and 21 days is significant higher than that in radiation control group. Hemoglobin (Hb) is not significantly different among these groups. Furthermore, the number of nucleated cells and CFU-GM colonies in bone marrow of DMC transplantation group are also significantly higher than that of in radiation control group on 15 days. Donor-origin cells could be detected in recipients on 15 days and 28 days after DMCs transplantation by Fluorescence in situ hybridization (FISH) analysis using Y-chromosome specific probes which indicates DMCs may play a direct cell-cell interaction and regulate the hematopoietic functions. In addition to this result, a question is that whether DMCs can repopulate the hematopietic system following lethal irradiation. The evidence shows that DMC transplantation can't rescue the animals following lethal irradiation of 8.5 Gy. Using a double-labeling technique to identify the lineage-specific markers, the Y-chromosome probe-positive transplanted cells don't express CD45 (a marker for hematopoietic progenitor cells) and indicate that DMCs are unable to differentiate into hematopoietic progenitor cells. By these results, multipotent cells isolated from dermis may at least represent an alternative origin of mesenchymal stem cells to restore marrow microenvironment and promote the survival, engraftment and proliferation of HSCs [38,39].

4. Plasticity of DMCs in Wound Healing

Implantation of DMCs into Skin Wounds

It has been well recognised that after skin injury new epidermis can derive from epidermal stem cells, but there is a debate in the origin of dermal fibroblasts and in the existence of dermal stem cells for long time. DMCs are identified from the dermal tissue that can generate multiple mesenchymal lineages including cartilage, muscle and surprisingly neuron-like cells. Although the origin of these multipotent stem cells is still unknown, how these stem cells participate in skin renewal and wound healing remains to be answered. A series of experiments are performed to study the potential functions of DMCs in wound healing. DMCs are labled using a fluorescent nuclear marker DAPI *in vitro* and are implanted into the minced skin. The subculture of labeled cells in vitro indicates that DAPI do not show obvious toxicity on the growth and division of the cells. After two weeks post-implantation when the tissue repair process is almost completed, cells in skin containing DAPI labeled nuclei are found in tissue sections under fluorescent microscope. From a comparative observation of the tissue sections under both light microscope and fluorescent microscope, the labeled cells distribute widely in dermis. Most of the labeled cells appear fibroblast-like morphology and differentiate into fibroblasts. Some labeled cells also differentiate into endothelial cells and are positive for factor VIII. Using a double-labeling technique to identify the lineage-specific markers of the positive cells, the transplanted cells are negative for cytokeratin (a marker for epidermal cells). Similar results are obtained using FISH technique with the Y-chromosome probe to distinguish the transplanted male cells from the female host. By these results, DMCs are able to engraft and differentiate into dermal cells in wounded skin. DNA microarray analysis has revealed that cultured DMCs express multiple cytokines and extracellular matrix genes that are associated with wound healing. Evidence further shows DMC supernatant can increase the proliferation of fibroblasts and epidermal cells that support DMCs can produce soluble products to promote wound healing. Macroscopic analysis of the wound reveal that topical implantation of DMCs requires a total period of 15 days for complete epithelialization, whereas the control groups take 18 days. Periodical observation of animals up to 14 days shows a significant increase in the rate of contraction of wounds in DMC-treated animals, when compared to control. This result indicates that DMCs could participate in skin regenerative process. There is increasing evidence to suggest that there are different types of fibroblasts in the skin, possibly with different functions.

Functional Regulations of DMCs by Wound Micorenvironment

Wound healing is a complex process including the network interactions of various kinds of cells, cytokines and extracellular matrices. Previous reports have indicated that wound fluid is a functional regulator of fibroblasts and endothelial cells. The effects of acute wound micorenviroment on DMCs are also observed. Wound fluid collected by polyvinyl alcohol (PVA) from the dorsum incision of normal adult rats is used in experiments. MTT assay shows that different concentrations of wound fluid from 5% to 30% increase the survival of cells, but over 10%, there is not obvious increase. Wound fluids significantly accelerate the migration rate of DMCs to fill in defects in monolayer cell sheet post-wounding in vitro. The hydroxyproline content in 10% wound fluid treated cells is higher than that in control by colorimetrical analysis indicate that DMCs differentiate into fibroblasts following treatment with wound fluids. To study the effects of in vivo wound microenvironment on DMCs, a method by using dialysis bags as carriers for cultured cells is established. DMCs are put into dialysis bags and are implanted into the dorsum of each animal through a midline incision. On day 3, when the early event (inflammatory process) for tissue repair is almost completes, the dialysis bags are taken out and the cells inside are observed under microscope and analysed. Immunohistochemistrial assay shows that bFGF and NGF express more strongly in cells from in vivo environment than that in cultured cells, while there is no obvious difference of bFGF and NGF receptors between DMCs in vivo and in vitro. However, the expression of a co-receptor for some growth factors, syndecan-1, is significantly increased in cells from the wound environment. In general, stem cells are quiescent in situ and are activated by the microenvironment after wounding [40]. In order to learn this process, DMCs are induced to enter a quiescent state by serum deprivation for 48 hours and then stimulated by addition of early-phase wound fluids. The intention is to use this experiment as a model to study the control of the transition of DMCs from G0 to a proliferationg state. Suppression subtractive hybridization (SSH) is used to screen the differentially expressed genes. The differentially expressed cDNA clones are further verified by sequencing and reverse Northern hybridization. The results show that the transcriptional response of DMCs to wound fluids is rapid and many genes with known roles in processes relervant to the physiology of wound healing are up-regulated. One stricking feture is that the expression of secretory leukocyte protease inhibitor (SLPI) increases in DMCs after stimulation by wound fluids. It is kown that SLPI inhibits chymotrypsin, trypsin, elastase, and cathepsin G. However, a positive correlation between cellular SLPI production and proliferation in DMCs is demonstrated. These results suggest that SLPI may define a novel pathway by which cellular growth of DMCs is modulated.

Hair Follicle Induction by DMCs

Skin substitutes/equivalent produced with cultured skin cells have been engineered for treatment of open skin in ulcers, wounds and burns. A major deficiency in all engineered skin is the lack of permanent appendant organs such as hair follicle, sweat glands and sebaceous glands. Multipotent stem cells provide an accessible source of target cells for skin tissue engineering with complex structutes. DMCs have been proved to can act as wound healing fibroblasts, a question is whether they can act as follicle dermal cells. When co-cultured with hair outer sheath cells (ORS) in vitro, DMCs can support the expansion of cytokeratin-19 positive ORS cells. Hair ORS cells can induce the expression of α-SMA (a marker for follicle dermal cells) in DMCs. Since α-actin is a marker for follicle dermal cells, this result suggests that intercellular interaction with hair outer sheath cells can promote the differentiation of DMCs. DMCs also can promote the growth of hair follicles in vitro. When the cell combination of DMCs and hair ORS epithelal cells is implanted subcutaneously in nude mice and new follicular-type structutes are formed in 60 days. Furthermore, an engineered skin equivalent is constructed with DMCs and hair ORS cells. The skin equivalent shows a coherent structure comprising a distinct dermis and epidermis separated by a continuous basement menbrane. Histological observation reveals normal characteristics of dermal and epidermal components. The dermal cells are fibroblastic and distributed in a collagen extracellular matrix. The epidermal cells have normal morphology and form epidermis with multile layers. Globally, these data shows DMCs may represent a source of dermal cells in wound healing and construction of complex skin equivakentsa nd skin substitutes [41].

5. Multiple Therapeutic Effects of DMCs in Combined Injury

It has been proved that DMCs can accelerate skin wound healing process in simple wound model and can promote hematopoitic recovery in sublethally irradiated model, respectively. A very interesting question is that whether DMCs can simultaneously reach multiple therapeutic effects. Combined radiation and wound injury (CRWI) is a typical combined injury which is very complex and difficult to deal with than single injuries. For the treatment of CRWI, to promote wound healing and hematopoietic recovery should be considered at the same time. Rats with CRWI are used in experiments. The survival rate of animals in 30 days in group received systemic transplantation of DMCs is 79%, whereas in the control group, the survival rate is only 53%. There is a marginally statistical difference between the two groups. The macroscopic analysis of the wound reveals that the DMC trasnplantation group heals faster than the control groups. Microscopic examination reveals that regenerating granulation tissue from the DMC transplantation group exhibit a highly accelerated granulation tissue formation relative to control rats. The densities of fibroblasts and blood vessels in granulation tissue are both significantly increased compared with control rats. Histological examination also shows that the epidermis in DMC transplantation group is thicker than control group. Systemic transplantation of DMCs also accelerates the blood recovery of rats with CRWI. The WBC counts in peripheral blood, the numbers of nucleated cell, CFU-GM and CFU-F colonies in bone marrow of DMC transplantation group recover earlier than that of control group. Using FISH analysis with the Y-chromosome probe to distinguish the transplanted male cells from the female host, exogenous cells are found in wounded recipient skin tissue and bone marrow after transplantation, suggesting that the implanted cells have engrafted and differentiated into target cells in the wound environments. It is further found that, following systemic infusion, DMCs distribute to a wide range of tissues, while distribute preferentially in the injuried tissues. By these data, systemic transplantation of multipotent stem cells is first proved to reach multiple therapeutic effects using a combined radiation and wound injury model. This wok provides a new strategy for the treatment of severe trauma and combined injuries [42].

6. Spontaneous and Malignant Transformation of DMCs

Although stem cells offer a potentially unlimited source of replacement cells or tissues for treating a myriad of diseases and disabilities, much still needs to be learned about stem cells, including their potential hazards of malignant transformation. The transformation risk of stem cells to produce tumours when expanded and transplanted could preclude their therapeutic usage. It has been described that stem cells and malignant cells share many characteristics. In many ways, stem cells and malignant cells also seem interchangeable. Recently, there is some evidence that some tumors, including malignant fibrous histiocytoma and osteosarcoma, may derive from undifferentiated mesenchymal stem cells. In order to learn the malignanat transformation risk of DMCs, clonal populations of rat and human DMCs are consecutively subcultured to maligant transformation in vitro. The transformed cells lose the cell-cell contact inhibition and grow in overlapping clusters in vitro. The doubling time of the transformed cells are about 24h when cultured in IMDM containing 10% FBS, whereas the parental DMCs are about 40h. The DNA histogram shows that the percentage for G0/G1 phase of transformed cells is 66%, while the percentage for G0/G1 phase of DMCs is 87%. The transformed cells can develop a detectable mass within 2 weeks after subcutaneous inoculation in nude mice, and tumors enlarge continuously until the mice die. The tumor produced by the transformed cells shows histological features of fibrous histiocytoma. Immunochemical study shows that most of the tumor cells are positive for the expression of vimentin, some cells are positive for S-100 and α-SMA. When the transformed cells are inoculated ectopically in the subcutis of nude mice, no metastatic lesions are observed macroscopically in the lungs and other organs. After induction with specific media, the transformed cells also exhibit multipotency and differentiate into cells with phenotypic properties of osteocytes and adipocytes.

Using Affymetrix microarray technology to compare expression profiles of 8784 genes in normal DMCs vs. spontaneously transformed cells, among the 8784 genes and ESTs analyzed, expression of 1057 shows to be altered by at least 2.0 fold in the transformed cells, of which 495 corresponded to ESTs or cDNA with unknown function. Among known genes, 352 are up-regulated, whereas 210 aere down-regulated after transformation. Some up-regulated genes are further confirmed by northern blotting analysis. These data show that, after transformation, most of genes encoding structural proteins, secretory proteins, receptors, proteases, protease inhibitors, extracellular matrix components, proteins involved in angiogenesis and cytosolic proteins, are down-regulated whereas genes encoding DNA-associated proteins (involved in DNA replication and reparation) and cell growth-related proteins are up-regulated. It is interesting to note that some genes encoding transcription factors, signaling proteins, membrane proteins and enzymes are up-regulated. Although we do not know how the transformation of DMCs occurs, these data may explain, at least in part, the behavior of transformed cells. The down-regulation of structural proteins, extracellular matrix components, secretory proteins and receptors is consistent with reversion of the phenotype of transformed cells towards a less differentiated phenotype and up-regulation of cell growth-related proteins and DNA-associated proteins is consistent with their accelerated growth. It is also interesting to note that approximately as many genes at least 4.0 fold (the signal log ratio ≥ 2) in transformed cells relative to the normal DMCs are up-regulated ($n = 74$) as down-regulated ($n = 71$). Among the oncogenes tested, c-jun mRNA for transcriptional factor AP-1(the signal log ratio =2.1), c-fos (the signal log ratio =1.9), c-ki-ras (the signal log ratio =1.9) and activated c-raf (the signal log ratio =1.1) are up-regulated. The up-regulated genes of cell growth-related proteins includes P38 mitogen activated protein kinase, MAP kinase kinase kinase 1(MEKK1), MAP-kinase phosphatase, GTP binding protein, GTPase activating protein, proliferating cell nuclear antigen (PCNA) and cyclin D1. It is also very interesting to note that these up-regulated oncogenes and cell growth-related proteins are all closely related to the Ras/Raf/MEK/ERK signalling pathway that play important roles in the control of cell proliferation and have strong transforming potential. The up-regulation of K-Ras gene is also confirmed by immunocytochemistry. The results also show that inhibition of ras/MEK pathway led to a significant decrease of the growth activity of the transformed cells in vitro and further support that the Ras/Raf/MEK/ERK signalling pathway may play a role in the malignant transformation of DMCs in vitro. [43,44]

7. Other Stem Cell Types in the Dermis

Skin stem cells are unipotent, meaning that they generate a single lineage, or multipotent, meaning that they generate multiple lineages. There are more stem cells in the dermis than the multipotent stem cells with mesenchymal potential. At least two different other stem cells (hematopoietic and neural stem cells) have been identified in the dermis.

Neural Crest Stem Cells

Neural crest stem cells (NCSCs) can be isolated from adult dermis and hair follicles. These stem cells, also referred as skin-derived progenitors (SKPs), are able to differentiate into neurons, nerve supporting cells, cartilage/bone cells, smooth muscle cells, and pigment cells [33,45]. These cells express nestin and fibronectin and can be passaged up to 12 months without senescence. Research shows that these stem cells can be obtained from different skins sites including the ear, the abdomen and the back. A major difference is that NCSCs can be cultured as non-adherent cells to form sphere colonies in vitro, while adherent mesenchymal stem cells can't proliferate under the same conditions. It is reported that EGF or bFGF is indispensable for the sphere formation, while TGF-β can increase the formation and the proliferation of skin-derived stem cell colonies without affecting their potency of differentiation. Adult neural stem cells from the skin with the addition of TGF-β during colony formation may become an accessible source for autologous cell-replacement therapy. When dermis-derived neural stem cells are transplanted into the spinal cord after traumatic injury, the cells can migrate to the lesion site, and there differentiate mainly in cells expressing glia and neuronal markers [46]. Cell replacement therapy shows particular promise inn the nervous system [47-49]. Embryonic stem cells can differentiate into any cell type of the body and have been shown to promote functional recovery in animal model [50]. However, their use raises ethical concerns because embryos are being destroyed in the process. In contrast, neural crest stem cells from adults have several advantages: they will not raise ethical concerns; they have the innate ability to differentiate into many diverse cell types; they are easily accessible in the skin of adults; and the patient's own neural crest stem cells could be used for cell therapy [51]. The latter avoids both rejection of the implant and graft-versus-host disease. In this regard, neural stem cells from the skin may provide a novel and powerful therapeutic strategy for the disease and trauma of nerve system.

Hematopoietic Stem Cells

Stem cells with mesenchymal potential have been identified from the follicle dermal cells and non-follicle dermis. These stem cells can act as a source of marrow stromal cells, while they can't differentiate into hematopoietic cells. However, recent evidence suggests that dermal compartments of the adult hair follicle have much broader stem cell activities than previously described. Recent research presents that mouse hair follicle end bulbs as well as micro-dissected dermal papilla and dermal sheath cells actively produce cells of erythroid and myeloid lineages. When cultured dermal papilla or dermal sheath cells are transplanted into lethally irradiated recipient mice, colony assays from bone marrow of primary recipients reveal that over 70% of clonogenic precursors are derived from donor hair follicle cells. When bone marrow from primary mice is harvested and used to repopulate secondary myeloablated recipients, multi-lineage haematopoietic engraftment is also observed [52]. Further research also shows that human dermal papilla contains a CD45 positive hematopoietic cell population. Cultured DP cells produce a small number of hematopoietic colonies (approximately 1 in 5000 DP cells plated) in CFU-GM assay. When cultured in the presence of interleukin-3 (IL-3, 20ng/ml), GM-CSF (10ng/ml) and granulocyte colony-stimulating factor (G-CSF, 20ng/ml) for seven days, the CD45 positive cells can be expanded about fifty times in cell number [53]. These data indicate that hair follicle may act as a source of extramedullary haematopoietic stem cell activity for blood stem cell therapeutics.

Acknowledgements

This work was supported by grants from National Science Foundation of China (NO. 30400188) and Ministry of Science and Technology of China (NO. 2005CB522605).

References

[1] Shi, C.M., Cheng, T.M., Su, Y.P. (2001). Exploring the developmental potential of adult somatic cells. *Chinese Science Bulletin, 46*, 4-5.

[2] Verfaillie, C.M. (2002). Adult stem cells: assessing the case for pluripotency. *Trends in Cell Biology, 12*, 502-508.

[3] Keller, R. (2002). Stem cells: on the way to restorative medicine. *Immunology Letters, 83(1)*, 1-12

[4] Owens, D.M. & Watt, F.M. (2003). Contribution of stem cells and differentiated cells to epidermal tumours. *Nat. Rev. Cancer, 3*, 444–451.

[5] Perez-Losada, J. & Balmain, A. (2003). Stem-cell hierarchy in skin cancer. *Nat. Rev. Cancer, 3*, 434–443.

[6] Friedenstein, A.J.P.-S.I. & Petrakova, K.V. (1966). Osteogenesis in transplants of bone marrow cells. *The Journal of Embryological Experimental Morphology, 16*, 381–390.

[7] Barry, F.P. & Murphy, M.J. (2004). Mesenchymal stem cells: clinical applications and biological characterization. *The International Journal of Biochemistry & Cell Biology, 36*, 568-584.

[8] Fukumoto, T., Sperling, J.W., Sanyal, A., Fitzsimmons, J.S., Reinholz, G.G., Conover, C.A. and O'Driscoll, S.W. (2003). Combined effects of insulin-like growth factor-1 and transforming growth factor-beta1 on periosteal mesenchymal cells during chondrogenesis in vitro. *Osteoarthritis Cartilage, 11*, 55–64.

[9] Tuli, R., Seghatoleslami, M.R., Tuli, S., Wang, M.L., Hozack, W.J., Manner, P.A., Danielson, K.G. and Tuan, R.S. (2003). A simple, high-yield method for obtaining multipotential mesenchymal progenitor cells from trabecular bone. *Molecular Biotechnology, 23*, 37–49.

[10] De Ugarte, D.A., Morizono, K., Elbarbary, A., Alfonso, Z., Zuk, P.A., Zhu, M., Dragoo, J.L., Ashjian, P., Thomas, B., Benhaim, P., Chen, I., Fraser, J. and Hedrick, M.H. (2003). Comparison of multi-lineage cells from human adipose tissue and bone marrow. *Cells Tissues Organs, 174,* 101–109.

[11] Zuk, P.A., Zhu, M., Mizuno, H., Huang, J., Futrell, J.W., Katz, A.J., Benhaim, P., Lorenz, H.P., Hedrick, M.H. (2001). Multilineage cells from human adipose tissue: implications for cell-based therapies. *Tissue. Eng, 7*, 211-228.

[12] De Bari, C., Dell'Accio, F., Tylzanowski, P. and Luyten, F.P. (2001). Multipotent mesenchymal stem cells from adult human synovial membrane. *Arthritis and Rheumatism, 44*, 1928–1942.

[13] Jankowski, R.J., Deasy, B.M. and Huard, J. (2002). Muscle-derived stem cells. *Gene Therapy, 9*, 642–647.

[14] Noort, W.A., Kruisselbrink, A.B., in't Anker, P.S., Kruger, M., van Bezooijen, R.L., de Paus, R.A., Heemskerk, M.H., Lowik, C.W., Falkenburg, J.H., Willemze, R. and Fibbe, W.E. (2002). Mesenchymal stem cells promote engraftment of human umbilical cord blood-derived CD34(+) cells in NOD/SCID mice. *Experimental Hematology, 30*, 870–878.

[15] Miura, M., Gronthos, S., Zhao, M., Lu, B., Fisher, L.W., Robey, P.G. and Shi, S. (2003). SHED: stem cells from human exfoliated deciduous teeth. *Proceedings of the National Academy of Sciences of the United States of America, 100*, 5807–5812.

[16] Brouard, M. & Barrandon, Y. (2003). Controlling skin morphogenesis: hope and despair. *Current Opinion in Biotechnology, 14*, 520-525.

[17] Chunmeng, S. & Tianmin, C. (2004). Skin : a promising reservoir for adult stem cell populations. *Medical Hypotheses, 62*, 683-688.

[18] Shi, C.M. & Cheng, T.M. (2001). Isolation and culture of multipotent stem cells derived from neonatal rat dermis. *Acta Academiae Medicine Militaris Tertiae, 9*, 1068-1070.

[19] Shi, C. & Cheng, T. (2003). Effects of acute wound environment on the neo-natal dermal multipotent cells. *Cells Tissues Organs, 175*, 177-185.

[20] Shi, C., Cheng, T., Su, Y. (2004). Isolation and identification of adult multipotent stem cells from dermis. *Infection Inflammation Repair, 5*, 158-161.

[21] Young, H.E., Steele, T.A., Bray, R.A., Hudson, J., Floyd, J.A., Hawkins, K., Thomas, K, Austin T, Edwards C, Cuzzourt J, Duenzl M, Lucas PA, Black AC Jr. (2001). Human reserve pluripotent mesenchymal stem cells are present in the connective tissues of skeletal muscle and dermis derived from fetal, adult, and geriatric donors. *Anat Rec, 264*, 51-62.

[22] Chunmeng, S., Tianmin, C., Yongping, S., Xinze, R., Yue, M., Jifu, Q., Shufen, L., Hui, X., Chengji, L. (2004). Effects of dermal multipotent cell transplantation on skin wound healing. *J Surg Res, 121*, 13-19.

[23] Shi, C.M. & Cheng, T.M. (2004). Differentiation of dermis-derived multipotent cells into insulin-producing pancreatic cells in vitro. *World J Gastroenterol, 10*, 2550-2.

[24] Shi, C. & Cheng, T. (2002). Long-term culture of dermis-derived multipotent stem cells and the effects of collagen sponge on their growth in vitro. *Sheng Wu Yi Xue Gong Cheng Xue Za Zhi, 19*, 660-663.

[25] Young, H.E., Duplaa, C., Romero-Ramos, M., Chesselet, M.F., Vourc'h, P., Yost, M.J., Ericson, K., Terracio, L., Asahara, T., Masuda, H., Tamura-Ninomiya, S., Detmer, K., Bray, R.A., Steele, T.A., Hixson, D., el-Kalay, M., Tobin, B.W., Russ, R.D., Horst, M.N., Floyd, J.A., Henson, N.L., Hawkins, K.C., Groom, J., Parikh, A., Blake, L., Bland, L.J., Thompson, A.J., Kirincich, A., Moreau, C., Hudson, J., Bowyer, F.P., Lin, T.J., Black, A.C. Jr. (2004). Adult reserve stem cells and their potential for tissue engineering. *Cell Biochem Biophys, 40*, 1-80.

[26] Lavker, R.M., Sun, T.T. (2000). Epidermal stem cells: properties, markers, and location. *Proc. Natl. Acad. Sci. U S A, 97*, 13473-13475.

[27] Morris, R.J., Potten, C.S. (1994). Slowly cycling (label-retaining) epidermal cells behave like clonogenic stem cells in vitro. *Cell. Prolif, 27*, 279-289.

[28] Turksen, K. and Troy, T. (1998). Epidermal cell lineage. Biochem. *Cell. Biol, 76*, 889–898.

[29] Oshima, H., Rochat, A., Kedzia, C., Kobayashi, K., Barrandon, Y. (2001). Morphogenesis and renewal of hair follicles from adult multipotent stem cells. *Cell, 104*, 233-245.

[30] Taylor, G., Lehrer, M.S., Jensen, P.J., Sun, T.T., Lavker, R.M. (2000). Involvement of follicular stem cells in forming not only the follicle but also the epidermis. *Cell, 102(4)*, 451-461.

[31] Jahoda, C. A. B. and Reynolds, A. J. (2000). Stem cells — a hairy issue. *Nat. Med, 6*, 1095-1097.

[32] Paus, R., and Cotsarelis, G. (1999). The biology of hair follicles. *N. Engl. J. Med, 341*, 491–497.

[33] Toma, J.G., Akhavan, M., Fernandes, K.J., Barnabe-Heider, F., Sadikot, A., Kaplan, D.R., Miller, F.D. (2001). Isolation of multipotent adult stem cells from the dermis of mammalian skin. *Nat. Cell Biol, 3*, 778-786.

[34] Fernandes, K.J., McKenziem, I.A., Mill, P., Smith, K.M., Akhavan, M., Barnabe-Heider, F., Biernaskie, J., Junek, A., Kobayashi, N.R., Toma, J.G., Kaplan, D.R., Labosky, P.A., Rafuse, V., Hui, C.C., Miller, F.D. (2004). A dermal niche for multipotent adult skin-derived precursor cells. *Nat Cell Biol, 6*, 1082-1093.

[35] Galotto, M., Berisso, G., Delfino, L., Podesta, M., Ottaggio, L., Dallorso, S., Dufour, C., Ferrara, G.B., Abbondandolo, A., Dini, G., Bacigalupo, A., Cancedda, R., Quarto, R.. (1999). Stromal damage as consequence of high-dose chemo/radiotherapy in bone marrow transplant recipients. *Exp Hematol, 9*, 1460-1466.

[36] Devine, S.M. & R. Hoffman. (2000). Role of mesenchymal stem cells in hematopoietic stem cell transplantation. *Curr Opin Hematol, 6*, 358-363.

[37] Koc, O.N., Gerson, S.L., Cooper, B.W., Dyhouse, S.M., Haynesworth, S.E., Caplan, A.I., Lazarus, H.M. (2000). Rapid hematopoietic recovery after coinfusion of autologous-blood stem cells and culture-expanded marrow mesenchymal stem cells in advanced breast cancer patients receiving high-dose chemotherapy. *J Clin Oncol, 2*, 307-316.

[38] Chunmeng, S., Tianmin, C. (2004). Effects of plastic-adherent dermal multipotent cells on peripheral blood leukocytes and CFU-GM in rats. *Transplant Proc, 36*, 1578-1581.

[39] Shi, C.M., Cheng, T.M., Su, Y.P., Mai, Y., Qu, J.F., Ran, X.Z. (2004). Transplantation of dermal multipotent cells promotes the hematopoietic recovery in sublethally irradiated rats. *Journal of Radiation Research, 45*, 19-24.

[40] Shi, C., Cheng, T., Su, Y., Mai, Y. Significance of dermis-derive dmultipotent stem cells in wound healing and skin equivalent construction. *Joint International Tissue Engineering Society (TESI) and European Tissue Engineering Society (ETES) Meeting, Lausanne October 10th-13th*, 2004, Switzerland

[41] Shi, C.M. & Cheng, T.M. (2003). Effects of wound environment on biological properties of dermal multipotent stem cells in rats. *Chin J Trauma, 19*, 199-202.

[42] Shi, C., Cheng, T., Su, Y., Mai, Y., Qu, J., Lou, S., Ran, X., Xu, H., Luo, C. (2004). Transplantation of dermal multipotent cells promotes survival and wound healing in rats with combined radiation and wound injury. *Radiat Res, 162*, 56-63.

[43] Shi, C.M., Cheng, T.M. (2001). Isolation and culture of multipotent stem cells derived from neonatal rat dermis. *Acta Academiae Medicine Militaris Tertiae, 9*, 1068-1070.

[44] Chunmeng, S., Tianmin, C., Yongping, S. (2004). Molecular alterations of dermis-derived multipotent stem cells after spontaneous and malignant transformation in vitro. *Journal of Tumor Marker Oncology, 19(S4)*, 23.

[45] Toma, J.G., McKenzie, I.A., Bagli, D., Miller, F.D. (2005). Isolation and characterization of multipotent skin-derived precursors from human skin. *Stem Cells, 23*, 727-737.

[46] Kawase, Y., Yanagi, Y., Takato, T., Fujimoto, M., Okochi, H. (2004). Characterization of multipotent adult stem cells from the skin: transforming growth factor-beta (TGF-beta) facilitates cell growth. *Exp Cell Res, 295*, 194-203.

[47] Joannides, A., Gaughwin, P., Schwiening, C., Majed, H., Sterling, J., Compston, A., Chandran, S. (2004). Efficient generation of neural precursors from adult human skin: astrocytes promote neurogenesis from skin-derived stem cells. *Lancet, 364*, 172-178.

[48] Belicchi, M., Pisati, F., Lopa, R., Porretti, L., Fortunato, F., Sironi, M., Scalamogna, M,, Parati, E.A., Bresolin, N., Torrente, Y. (2004). Human skin-derived stem cells migrate throughout forebrain and differentiate into astrocytes after injection into adult mouse brain. *J Neurosci Res, 77*, 475-486.

[49] Gorio, A., Torrente, Y., Madaschi, L., Di Stefano, A.B., Pisati, F., Marchesi, C., Belicchi, M., Di Giulio, A.M., Bresolin, N. (2004). Fate of autologous dermal stem cells transplanted into the spinal cord after traumatic injury (TSCI). *Neuroscience, 125*, 179-189.

[50] McDonald, J.W., Liu, X.Z., Qu, Y., Liu, S., Mickey, S.K., Turetsky, D., Gottlieb, D.I., Choi, D.W. (1999). Transplanted embryonic stem cells survive, differentiate and promote recovery in injured rat spinal cord. *Nat Med, 5*, 1410-1412.

[51] Bjorklund, A., Lindvall, O. (2000). Cell replacement therapies for central nervous system disorders. Nat. Neurosci, 3, 537-544.

[52] M. Lako, L. Armstrong, P.M. Cairns, S. Harris, N. Hole and C. A. B. Jahoda. (2002). Hair follicle dermal cells repopulate the mouse haematopoietic system. *Journal of Cell Science, 115*, 3967-3974.

[53] Shi, C., Mai, Y., Cheng, T. (2004). Identification of hematopoietic cell populations from the dermal papillae of human hair follicles. *Transplantation Proceedings, 36*, 3208-3211.

In: Frontiers in Stem Cell Research
Editor: Julia M. Spanning pp. 153-171
ISBN 1-60021-294-8

Chapter VI

STEM CELL THERAPY FOR TYPE 1 DIABETES MELLITUS

Christopher J. Burns[1,*], Helen Milne[2], Peter M. Jones[2] and Shanta J. Persaud[2]

[1]Division of Immunology and Endocrinology, National Institute for Biological Standardisation and Control, Potters Bar, Hertfordshire, EN6 3QG, UK;
[2]Beta Cell Development and Function Group, Division of Reproductive Health Endocrinology and Development, King's College London, London, SE1 1UL, UK.

ABSTRACT

Type 1 diabetes mellitus (T1DM) is a disorder of a single cell type, the insulin-secreting pancreatic β-cell, and as such is particularly amenable to treatment by cell replacement therapy. For this reason, T1DM has received much attention recently as a potential target for the emerging science of stem cell medicine. In this autoimmune disease, the pancreatic β-cells are selectively and irreversibly destroyed by autoimmune assault. Advances in islet transplantation procedures now mean that individuals with T1DM can be cured by human islet transplantation. A major drawback in this therapy is the limited availability of donor islets, and the search for substitute transplant tissues has intensified in the last few years. This review will summarise recent progress in using stem cell populations for generating substitute β-cells for transplantation therapy. The relative merits of stem cells from different sources will be considered, and the major technical obstacles to the production of reliable and reproducible stem cell to β-cell differentiation protocols will be addressed.

Keywords: Stem cells, diabetes mellitus, pancreas, beta-cell, islets of Langerhans

* Correspondence concerning this article should be addressed to Dr Chris Burns, National Institute for Biological Standardisation and Control, Potters Bar, Hertfordshire, EN6 3QG, UK

1. Introduction

Type 1 diabetes mellitus (T1DM) is a single cell disorder in which pancreatic β-cells are irreversibly destroyed by autoimmune assault. Since the discovery of insulin in the 1920s, the main therapeutic approach to insulin-dependent diabetes has been to treat the symptoms by insulin replacement. However, in many cases individuals with T1DM still do not achieve appropriate glycaemic control and the abnormal blood glucose levels lead to a high risk of long-term complications. Recent advances in transplantation and immunosuppression techniques suggest that transplantation of human primary islets of Langerhans may offer a cure for T1DM (Shapiro *et al.*, 2000; Ryan *et al.*, 2001; Gaglia *et al.*, 2005).

Human islet transplantation has produced impressive results, with some patients maintaining insulin independence for a number of years, but the success of this therapeutic intervention requires islets harvested from up to four pancreata from heart-beating donors for each recipient. The islet transfer to each recipient is equivalent to approximately $1x10^6$ primary human islets (~$2x10^9$ β-cells) and so it is clear that very large numbers of replacement β-cells will be required to make a wide-spread therapeutic impact on T1DM (up to ~10^5 potential recipients in the UK; ~10^6 in the USA).

The need for the large numbers of β-cells required could be met from a substantial supply of non-human islets: using primary islets from other species as transplant material is an obvious strategy, but one that has not made substantial progress despite considerable efforts. Pigs have been used for much of this work since high islet yields can be readily obtained and they are amenable to genetic modification, but there are important impediments to using pig islets for transplantation therapy of T1DM in humans. Thus, although the hyperimmune response to xenografts can be limited by islet encapsulation, this has largely failed to maintain islet viability and insulin secretory responses over the prolonged periods required of transplantation therapy. Another problem associated with xenografts is the possibility that retroviral sequences in the porcine DNA may become activated on xenotransplantation (van der Laan *et al.*, 2000), which could result in the generation of new viral infections in humans receiving porcine islet implants.

Given the potential drawbacks to the use of xenografts described above, it is highly unlikely that xenografted β-cells will find any major clinical application until these problems are resolved. This review will therefore focus on the possibilities of deriving replacement β-cells from a proliferative precursor population(s) that can be expanded considerably *in vitro* before differentiation into β-cells. Stem cells of adult or embryonic origin are obvious candidates because of their ability to replicate to the numbers required, and their potential to differentiate into a range of tissue types including β-cells. There is currently no consensus on the most suitable type of progenitor material, but a number of potential options are available, including multipotent adult stem/progenitor cells from a range of tissues, and pluripotent embryonic stem (ES) cells.

2. Functional Properties of Replacement β-Cells

Any tissue destined for transplantation therapy of T1DM will need to have some essential functional abilities. Most important of these will be the insulin secretory function of the replacement cells that is required for maintaining normoglycaemia *in vivo*. The replacement cells must therefore be able to synthesise and store sufficient insulin to maintain normoglycaemia, and to release appropriate amounts of insulin in response to changes in the ambient glycaemia. Pancreatic β-cells have a complex stimulus-response coupling mechanism that enables them to monitor and respond rapidly to changes in circulating nutrients and a variety of hormones and neurotransmitters. These mechanisms have been defined in some detail (Figure 1; reviewed by Jones and Persaud, 1998) and substitute β-cells will have to express the appropriate functional elements to permit a regulated release of insulin in response to the appropriate environmental cues. This is an important point in an area of research in which the ability of a cell to express the preproinsulin gene is often taken as a valid end-point in differentiation protocols.

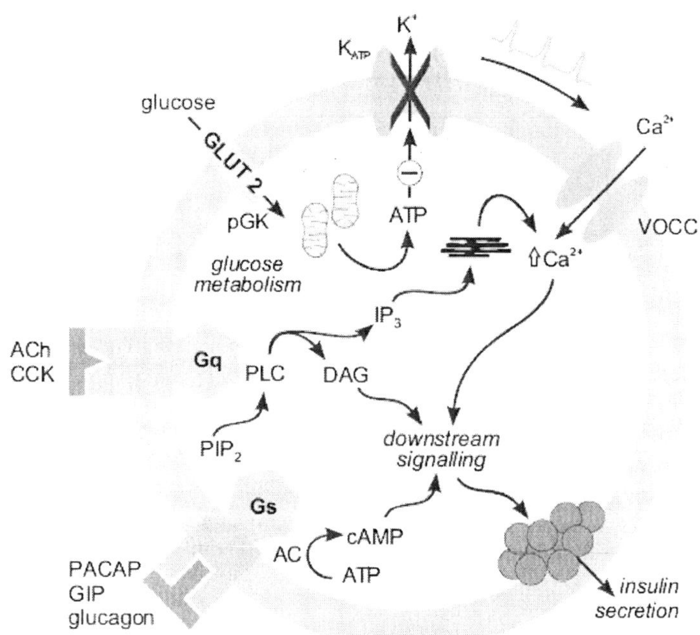

Figure 1. Signal transduction in primary β-cells. Glucose is transported into β-cells on the high capacity GLUT2 transporter and phosphorylated by the pancreatic form of glucokinase (pGK). ATP that is generated following glucose metabolism binds to ATP-sensitive K^+-channels (K_{ATP}), leading to closure of the pore-forming Kir6.2 channel subunits. The consequent inhibition of K^+ efflux leads to β-cell plasma membrane depolarisation and opening of voltage-operated Ca^{2+} channels (VOCC). Extracellular Ca^{2+} enters β-cells through the open VOCCs, resulting in elevations in intracellular Ca^{2+}, activation of Ca^{2+}-sensitive downstream signalling pathways and initiation of insulin secretion. Insulin secretion is also stimulated by ligands such as acetylcholine (ACh) and cholecystokinin (CCK) acting at cell surface receptors linked to IP_3 and DAG generation and those such as pituitary adenylate cyclase activating polypeptide (PACAP), glucose-dependent insulinotropic peptide (GIP) and glucagon, which elevate intracellular cyclic AMP.

A second important issue is the proliferative capacity *in vivo* of the replacement β-cells. Proliferation of transplanted primary β-cells derived from organ donors is not a problem since they have a low proliferative capacity, but the use of replacement β-cells that are generated from proliferative precursors such as stem cells may present problems. Under these circumstances, incomplete control of post-transplantation proliferation would result in the undesired development of hyperinsulinaemic hypoglycaemia following expansion of β-cell mass *in vivo*.

It is obviously vital that transplanted cells are not destroyed by the recipient's immune system, and, currently, if individuals were to receive insulin-expressing cells derived from starting material such as human ES (hES) cells or tissue stem cell populations they would still have to undergo a lifetime of immunosuppression. To address this, attempts are being made to circumvent the immune response by using a variety of non-immunosuppressive strategies. Thus, the generation of hES cells from blastocysts derived by somatic cell nuclear transfer (SCNT; (Hwang *et al.*, 2005) could lead to individuals with T1DM receiving a transplant of immunologically matched, hES cell-derived, insulin-expressing cells. In addition, differentiated mouse ES (mES) cells derived by nuclear transfer have been used successfully to treat Parkinson's disease in a mouse model (Barberi *et al.*, 2003). Another means of transplantation therapy for diabetes without the need for immunosuppression would be through trans-differentiation of bone marrow stem cells into insulin-expressing cells. Bone marrow-derived cells are widely used to treat a variety of haematological diseases and the transplantation of autologous bone marrow-derived cells in a patient with diabetes could prevent graft rejection. However, even immunologically autologous β-cell replacements would still be targeted by the immune system of an individual with T1DM since it is programmed to recognise and destroy primary β-cells. It is possible that this autoimmune destruction of β-cells could be avoided by generating insulin-secreting cells that show the functional characteristics of primary β-cells but that are developmentally and immunologically distinct. In this way they could evade the immune assault without immunosuppression, and this possibility is addressed below.

3. Sources of Replacement β-Cells

The proliferative capacity and pluripotency of stem cells make them attractive candidates for the generation *in vitro* of insulin-expressing β-cells for transplantation therapy *in vivo*. There are several different types of progenitor cell populations and there is currently no clear consensus on which, if any, of these populations will prove to be therapeutically useful. Possible progenitor cells are considered here under two categories - the broad category of tissue progenitor cells that are normally found in some adult and fetal tissues, and the more specific category of ES cells that are generated *in vitro* from blastocysts.

a. Tissue Progenitor/Stem Cells

The availability of β-cells for transplantation could be increased if the populations of progenitor or stem cells that are present in a variety of tissues could be identified, isolated and expanded *in vitro*, and then induced to differentiate into a β-cell phenotype. Tissue stem cells are usually considered to be lineage-restricted to mature into the differentiated cells of the host tissue, but evidence is emerging of trans-differentiation into other types of tissue, suggesting that several different tissues may offer sources of progenitor cells that have the potential to become insulin-expressing cells.

i. Pancreas/ Islet Progenitors

Stem cells that can be induced to express markers of a β-cell phenotype are thought to exist in the adult pancreas (Bonner-Weir et al., 2000; Ramiya et al., 2000; Zulewski et al., 2001; Abraham et al., 2002) although the location and identity of these progenitor cells have not been unequivocally established (Lechner and Habener 2003). It has been suggested that pancreatic ductal cells, which retain some ability to de-differentiate and acquire elements of the β-cell phenotype, may be the progenitors for β-cell neogenesis (Bonner-Weir et al., 2000; Abraham et al., 2002). Islets contain cells that express the intermediate filament protein nestin, often considered as a stem cell marker (Zulewski et al., 2001; Hunziker and Stein 2000), so it is possible that β-cell neogenesis may also occur within the islets themselves. Nestin-positive pancreatic cells can be induced to express β-cell markers (Zulewski et al., 2001), but several studies question the use of selection of nestin-positive cells as β-cell precursors. Thus, nestin is thought not to be expressed in progenitor cells during the development of mouse (Sealander and Edlund, 2002) or human (Hanley 2002) pancreas, and it has been reported that islet nestin-positive cells are not associated with the endocrine compartment (Klein et al., 2003; Street et al., 2004) and that nestin-positive cells from fetal human pancreas do not express β-cell markers (Humphrey et al., 2003).

An alternative method of generating β-cell populations from islet progenitors is to utiliise epithelial-to-mesenchymal transition (EMT), which has been shown to occur during development and carcinogenesis (Hay, 1995; Thiery, 2002; 2003). It has been reported that human islets maintained in culture produced proliferative fibroblastic mesenchymal-type cells that arose through the de-differentiation of islet epithelial cells by EMT, and these cells could re-differentiate into insulin-expressing cells (Gershengorn et al., 2004). Although potentially promising, it is difficult to fully characterise the potential of this experimental approach in the absence of clonal isolation and proliferation of individual precursor cells. A recent study addressed this using a colony-forming assay in which pancreas precursors were identified in both nestin-positive and nestin-negative pancreatic fractions, and they showed glucose-dependent calcium responses and insulin release (Seaberg et al., 2004). Although the studies outlined above support the possibility of β-cell progenitors existing within the pancreas, recent cell-labelling studies in transgenic mice have suggested that the expansion in β-cell mass that occurs during normal growth and development can be entirely attributed to proliferation of pre-existing β-cells and is independent of any pancreatic stem cells (Dor et al., 2004). There is currently no evidence to suggest that β-cells of the adult human pancreas have anything more than a limited proliferative potential or are amenable to the expansion required to produce cells in the large quantities required for transplantation therapy.

ii. Liver Progenitor Cells

Progenitor cells located within tissues other than the pancreas may also be suitable as a starting material in cell replacement studies. In the search for a "non-pancreatic progenitor cell", the liver is an obvious starting point for a number of reasons. Firstly, the liver is relatively accessible for biopsy and has the capacity to regenerate and so would be a clinically attractive tissue for potential autologous grafting of liver-derived cells in patients with T1DM. Secondly, liver and pancreas are phenotypically similar in that they both develop from an endodermal origin in the embryo and share the expression of several phenotype-maintaining transcription factors, such as the hepatic nuclear factors (Odom et al., 2004). Thirdly, there are some important functional similarities between hepatocytes and β-cells. Thus, both cell types monitor changes in circulating glucose concentrations by the expression of the high capacity glucose transporter GLUT2, and in both cell types the cellular entry of glucose is accompanied by its rapid phosphorylation by a high specificity, low affinity glucokinase. Transdifferentiation is a phenomenon that describes the conversion of one cell type into another and there is some evidence that hepatic cells can undergo transdifferentiation into a β-cell phenotype. A likely candidate for the master regulator of this event in hepatic cells is the transcription factor PDX-1 (pancreatic duodenum homeobox-1), which is required for pancreas development and for the maintenance of a mature β-cell phenotype (Scharfmann, 2000). Thus, there have been a number of reports suggesting that the expression of PDX-1 in hepatic cells will generate cells with many functional aspects of pancreatic β-cells (Ferber et al., 2000; Horb et al., 2003; Kojima et al., 2003; Zalzman et al., 2003; Sapir et al., 2005; Li et al., 2005). Other reports have suggested that oval/stem cells from rodent liver can be induced to express β-cell markers in vivo (Suzuki et al., 2002) and in vitro (Yang et al., 2002), without the ectopic expression of transcription factors such as PDX-1. Although a sub-population of cells within the liver may have the capacity to transdifferentiate into cells expressing some elements of a β-cell phenotype, there is little evidence to date that they are amenable to the in vitro expansion that would be required for an effective transplantation therapy.

iii. Bone Marrow-Derived Stem Cells

The possibilities presented by bone marrow-derived stem cells are attractive since these cells are already routinely collected from adults in a clinical environment and have the distinct advantage that they could be used as an autograft. Bone marrow-derived stem cells have the potential to differentiate into a wide variety of tissue and cell types (Asahara et al., 1999, Ferrari et al., 1998, Pereira et al., 1995, Kopen et al., 1999, Schwartz et al., 2002) and there have been several recent reports that they may also act as β-cell precursors. Thus, maintenance of rat and mouse bone marrow cells in culture can generate insulin-positive cells (Jahr and Bretzel, 2003; Tang et al., 2004). In addition, recent studies in vivo using genetically-tagged adult mouse bone marrow cells (Ianus et al., 2003) or human cord blood cells (Yoshida et al., 2005) demonstrated that they engrafted into pancreatic islets and started to produce insulin. Although in some studies (Ianus et al., 2003; Tang et al., 2004) it has been suggested that the insulin-expressing cells are generated by trans-differentiation rather than the fusion of donor stem cells to host cells, one recent study (Yoshida et al., 2005) suggests that both fusion-dependent and –independent events take place. The identity of the bone marrow stem cells that can fuse with, or transdifferentiate into cells with a β-cell phenotype is unknown. Haematopoietic stem cells show a very low frequency of trans-differentiation into endodermal cells (Wagers et al., 2002), so it is more likely that mesenchymal stem cells are the precursors, and this is consistent with the demonstration that human bone marrow mesenchymal stem cells can be induced to express the insulin gene (Moriscot et al., 2005). However, the derivation of insulin-expressing cells from bone marrow stem cells is a controversial area and there have been reports of no significant in vivo differentiation of bone marrow into insulin-expressing cells in mice, although donor bone marrow did contribute to other cell types within the islet (Choi et al., 2003; Lechner et al., 2004). It is clear that it is still too early to assess whether bone marrow will be useful for the generation of expandable β-cell precursor populations, and much work still needs to be done in this area.

iv. Neural Progenitor Cells

Central nervous system-derived stem cells proliferate in culture for prolonged periods (Minger et al., 1996) and, although they are ectodermal in origin, can differentiate into cells of all three germ layers (Clarke et al., 2000; Sang et al., 2002). The use of neural progenitor cells as β-cell precursors is an interesting possibility since β-cells and neurons have many common characteristics: both are specialised cell types that synthesise signalling molecules that are stored in secretory vesicles and released by regulated exocytosis. In addition, sub-populations of neurons are known to express many functional elements that are characteristic of β-cells (see Figure 1), including ATP-dependent K^+ channel SUR1 and Kir6.2 subunits, voltage-operated Ca^{2+} channels, GLUT 2 glucose transporters and the pancreatic form of glucokinase (Yang et al., 1999). Furthermore, β-cell and neuronal development pathways show similarities (Figure 2), with many transcription factors (e.g. PDX-1, nkx2.2, nkx6.1, neurogenin-3, NeuroD1/Beta2, Pax4, Pax6 and Isl1) regulating both processes (Wilson et al., 2003; Burns et al., 2005; Hori et al., 2005). We have recently reported that rat neural stem cells can be differentiated in vitro towards an insulin-expressing phenotype (Burns et al., 2005), and another recent report has shown that glucose-responsive insulin-expressing cells can be derived from a transformed human neural cell line (Hori et al., 2005). It is not clear from these studies if endodermal β-cells are being derived from neuroectodermal starting populations, but, given that neural stem cells express nestin, the possibility exists that rather than trans-differentiation taking place β-cell elements are being expressed by the neuroectodermal cells. If this is the case, it may be advantageous for transplantation therapy to generate insulin-secreting neural cells that are more likely than authentic β-cells to avoid the autoimmune β-cell destruction that occurs in individuals with T1DM.

b. Embryonic Stem Cells

Embryonic stem (ES) cells are considered to have great therapeutic potential since their pluripotent nature means that cells from all three embryonic germ layers can be derived from them. ES cells are obtained from the inner cell mass (ICM) of blastocysts and can proliferate indefinitely *in vitro* in the presence of leukemia inhibitory factor (LIF) and/or embryonic fibroblast feeder cells. The conditions for mouse ES (mES) cell maintenance and expansion in culture are far simpler than those for human ES (hES) cells, and mES cells are not subject to the ethical and legal constraints that accompany hES cells. For these reasons most experimental studies to date have used mES cells.

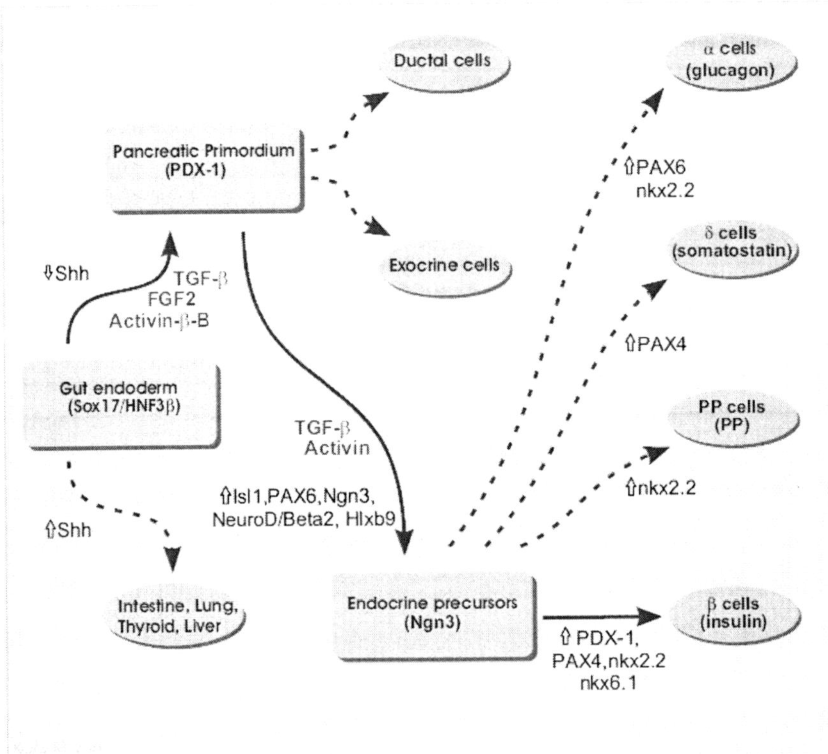

Figure 2. Regulation of pancreatic differentiation by transcription factors. Transcription factors such as transforming growth factor-β (TGF-β), fibroblast growth factor-2 (FGF2) and activin-β-B repress sonic hedgehog (Shh) signalling and this, coincidental with induction of pancreatic duodenal homeobox transcription factor-1 (PDX-1) expression, causes the formation of the pancreatic primordium from the Sox17/hepatocyte nuclear factor-3β-expressing gut endoderm. In the presence of TGF-β and activin the pancreatic primordium gives rise to neurogenin-3 (Ngn3)-expressing endocrine precursor cells and the subsequent formation of the islet endocrine cells is controlled by a complex network of transcription factors including NK-homeodomain transcription factors (nkx2.2 and nkx6.1) and the paired homeobox transcription factors (Pax4 and Pax6). Solid arrows denote the differentiation pathway towards a mature pancreatic β-cell and dashed arrows denote the pathways towards other cell phenotypes.

i. Mouse Embryonic Stem Cells

The exciting report that mouse ES cells could be induced to adopt some elements of the β-cell phenotype, most notably that they contained and released significant amounts of insulin (Lumelsky et al., 2001), provided a stimulus for numerous groups worldwide to reproduce a protocol that had originally been designed for neural stem cells. The basis of the differentiation procedure was to select for nestin-expressing cells in differentiating populations of mES cells growing as three-dimensional clusters known as embryoid bodies (EBs). There have been several subsequent independent reports suggesting that mES cells can differentiate into an insulin-expressing phenotype following LIF withdrawal and growing the cells as EBs, followed by enrichment of endocrine progenitor cells by culturing under selective conditions (Hori et al., 2002; Shiroi et al., 2002; Sipione et al., 2004; Bai et al., 2005). However, it is likely that most of the insulin detected in the original study, and some of the later ones, had been taken up from the extracellular medium (Rajagopal et al., 2003), with perhaps only a small component originating from synthesis de novo (Paek et al., 2005). Furthermore, a review of the protocols used for pancreatic cell derivation from mES cells suggests that selection of nestin-expressing cells results in neural differentiation and the authors suggest that such protocols should be avoided if the ultimate aim is production of insulin-producing cells (Kania et al., 2004). This caution is well-advised, and unless major advances are made in differentiation protocols of neural progenitor cells (see section 3a iv above) there seems to be little strategic advantage in using nestin positive cells as β-cell precursors.

An alternative approach to the use of selection protocols is to genetically manipulate ES cells by overexpressing transcription factors that regulate β-cell development, to drive their differentiation towards a β-cell fate (Figure 2). Some success has been achieved through the overexpression of Pax4 (Blyszczuk et al., 2003), PDX-1 (Miyazaki et al., 2004) and nkx2.2 (Shiroi et al., 2005), all of which have induced differentiation of mES cells towards a β-cell phenotype. However, the process of pancreatic development in vivo is complex and it is unlikely that permanent overexpression of single factors in vitro will recapitulate the developmental cascade. Additional progress is likely to be made in this area when further information is available on the timing and extent of transcription factor expression that is required to direct the development of mES cells to mature β-cells,

ii. Human Embryonic Stem Cells

As outlined above, there have been several publications on the derivation of insulin-expressing cells from mES cells in the past 5 years, but any cells destined for transplantation therapy for T1DM would obviously need to be derived from hES rather than mES cells. Human and mouse ES cells differ in some key elements and it is not clear whether information provided by mES cell studies will be useful when applied to hES cells. The major drawbacks to using hES cells is that most lines require a feeder layer of fibroblast cells, their growth characteristics as individual colonies means that they require labour-intensive physical dissection, and they tend to spontaneously differentiate in an unregulated manner. These technical difficulties, together with the ethical constraints associated with hES cell use and there being relatively few hES cell lines widely available for experimental use, might account for the paucity of reports of generating insulin-expressing cells from hES cells. When hES cells do spontaneously differentiate, one of the cell types produced are those of a β-cell phenotype. However, this occurs at very low frequency, with only 1-3% of cells expressing insulin (Assady et al., 2001; Segev et al., 2004; Burns et al., 2004). Another recent report has found that hES cells can spontaneously differentiate into pancreatic progenitors, but only express insulin when transplanted in vivo with mouse embryonic dorsal pancreas (Brolen et al., 2005). It thus seems likely that although there may be some limited capacity for hES cells to differentiate spontaneously into insulin-expressing cells, appropriate synthesis and processing of insulin requires extracellular cues from the developing pancreas.

4. Obstacles to Generating β-Cells from Stem Cells

This has proven to be a rapidly-evolving field of research which has produced a substantial number of published studies over the past few years. It is perhaps disappointing that, to date, none of the published protocols reports a method that is guaranteed to reproducibly generate large numbers of functional β-cells and there remains considerable confusion about how best to differentiate various stem cell populations into functional β-cells. There are many potential reasons for this lack of reproducibility in complex, multi-variate experimental protocols and it is worth considering briefly some of the possible sources of variability between studies.

a. Starting Material

Numerous different ES cell lines are currently in use and it is likely that some of the differences in experimental results reflect differences in the phenotype and the *in vitro* developmental potential of these lines. Anyone with experience of growing mouse or human ES cells *in vitro* will know that each cell line has its own environmental requirements (*e.g.* cell density, culture substrate, medium, growth factors, passage requirements) for continued proliferation and maintenance in the undifferentiated state, which implies pre-existing phenotypic differences between ES cell lines. More importantly, data contained in published studies attempting to generate β-cells from mES cells suggest that significant variations may exist between the starting ES cell populations (see Table 1). For example, Hori and colleagues (Hori et al., 2002) noted that the JM1 cell line consistently produced cells with higher levels of insulin expression than did the ROSA cell line. Similarly, the authors of another report (Lumelsky et al., 2001) commented on the expression of PDX-1 mRNA in their undifferentiated ES cell starting material, stating that in the absence of feeder cells E14.1 mES cells readily expressed the PDX-1 gene, perhaps suggesting a predisposition of this cell line to favour endoderm formation over ectoderm or mesoderm. This is not an isolated observation because comparison of PCR analyses from a number of β-cell differentiation papers (Table 1) demonstrates that a range of markers associated with pancreatic development (Figure 2) were already expressed in the proliferative ES cells prior to the differentiation protocols. These include PDX-1 (Kim et al., 2003; Houard et al., 2003; Lumelsky et al., 2001), NGN3 (Houard et al., 2003; Hori et al., 2002; Moritoh et al., 2003; Kahan et al., 2003) GLUT 2 (Lumelsky et al., 2001) and nestin (Hori et al., 2002). The presence of NGN3 and PDX-1 transcription factors are particularly interesting since they both play key roles in pancreatic development. PDX-1 is expressed in the region of the gut that is patterned for pancreas development, and is essential for normal formation of the pancreas (Ahlgren et al., 1996). NGN3 plays an equally crucial role, promoting specialisation of endocrine progenitor cells in the pancreas (Apelqvist *et al.,* 1999). Since both these factors act at an early developmental stage their expression could indicate the presence of pancreatic precursor cells, which, if cultured under conditions favourable for their survival, expansion and maturation, could result in differentiation to islet cells. The transition to primitive endodermal cells, the first differentiation event for the ICM, may also begin earlier than previously thought. In mouse development the primitive endoderm appears on the surface of the ICM at around embryonic day 4.5, as a layer of cells distinct from the inner epiblast layer which will eventually form the embryo, but there is evidence that commitment to this lineage may occur earlier, in individual cells within the ICM which are then segregated to the surface (Rossant, 2004). This has the obvious implication that ES cell lines derived from the ICM may already be committed to the primitive endoderm lineage.

Table 1. Mouse ES cell lines used for beta cell differentiation studies expressed genes associated with pancreatic development

Gene	Cell line	Ref.
PDX-1	R1[2] , SV-129 or C57/BL6[1] E14.1[3]	(Kim et al[1]., 2003; Houard et al[2]., 2003; Lumelsky et al[3]., 2001)
NGN-3	R1[2,7], JM1[4], D3[6], EB3[5]	(Houard et al[2]., 2003; Hori et al[4]., 2002; Moritoh et al[5]., 2003; Kahan et al[6]., 2003; Blyzsczuk et al[7], 2003)
GLUT-2	E14.1	(Lumelsky et al[3]., 2001)
nestin	JM1[4]	Hori et al[4]., 2002

b. Differentiation Protocols

An important source of variability between studies is the use of cell-cell interactions to drive differentiation. The cell-cell contact between ES cells that is induced by EB formation introduces an uncontrolled stage in the differentiation process that tends to generate very low numbers of insulin-containing cells within a phenotypically mixed and unpredictable cell population (Abe *et al.*, 1996). The alternative strategy of post-EB genetic selection of insulin-expressing cells (Soria *et al.*, 2000, Moritoh *et al.*, 2003) or Nkx6.1-positive cells (Leon-Quinto *et al.,* 2004) does not solve the problem of how to differentiate mES cells towards this phenotype, but merely selects for those cells that have spontaneously adopted the insulin-expressing phenotype, often as a consequence of EB formation. Recent studies have been directed towards developing simpler and more controlled differentiation protocols to more reproducibly generate endodermal insulin-expressing cell populations. One strategy involves bypassing embryoid body (EB) formation and culturing mES cells as a monolayer at high density in the continued presence of differentiation-suppressing factors to restrict the lineage of the resultant cell populations (Milne *et al.*, 2005).

Another source of variation between studies is in the use of biologically active factors, individually or in cocktail, to enhance the differentiation of the precursor population. For example, recent publications using mES cells have reported the successful generation of early endocrine pancreas using nicotinamide and exendin-4 (Ku *et al.,* 2004), the generation of pancreatic endoderm using all-*trans* retinoic acid (Micallef *et al.*, 2005) or pancreatic β-cells using Activin A and all-*trans* retinoic acid (Shi *et al.*, 2005).

c. Differentiated Cell Phenotype

Many current studies, particularly those using ES cells, attempt to use information derived from studies of pancreatic development (see Figure 2) to inform their *in vitro* differentiation protocols. However, it may not prove possible to recapitulate *in vitro* the complex sequence of developmental events *in vivo* directing gut endoderm towards an insulin-expressing β-cell phenotype, and there is some evidence that it may be possible to arrive at an insulin-expressing phenotype without following the normal pathway of islet development. Thus, it has been suggested that the pathways of β-cell differentiation *in vitro* may differ significantly from those *in vivo* (Houard *et al.*, 2003) and mES cells are reported to differentiate into insulin-expressing cells without PDX-1 expression (Moritoh *et al.*, 2003), although this is essential *in vivo* (Scharfmann 2000).

This raises the real possibility that the insulin-expressing cells produced by some *in vitro* differentiation protocols are not authentic endodermal, pancreatic β-cells, and in the absence of specific and highly-expressed markers for authentic β-cells it is difficult to determine unambiguously the origin of insulin-expressing cells generated *in vitro* from pluripotent progenitor populations. One alternative developmental origin for insulin-expressing cells could be the visceral endoderm of the yolk sac (Rau *et al.*, 1989; McGrath *et al.*, 1997). Visceral endoderm develops initially in a similar manner to embryonic endoderm, and primitive endodermal cells are localised to the periphery of EBs during differentiation of mES cells (Murray *et al.*, 2001). It has therefore been suggested that the insulin-expressing cells generated during EB formation are from visceral endoderm, since they are located almost exclusively in the outer layer of the EB (Houard *et al.*, 2003). Another alternative identity for the insulin-expressing cells may be neuroectodermal cells that are essentially neuronal but that express insulin and β-cell stimulus-response elements. This identity would be consistent with reports of transient expression of preproinsulin in the fetal brain during development (Devasker *et al.*, 1993ab), and of the expression of β-cell functional elements (see Figure 1) in a sub-population of hypothalamic neurons (Yang *et al.*, 1999).

For therapeutic purposes the precise developmental identity of the transplanted cells may not be important so long as they are available in sufficiently high numbers and fulfil the functional criteria of replacement β-cells, as described above. Indeed, for therapeutic purposes it may be advantageous to generate insulin-secreting cells that are not authentic β-cells since they may avoid the β-cell autoimmune destruction seen in individuals with T1DM.

5. Conclusions

Stem cells offer new therapeutic possibilities for the treatment of diabetes and the potential gains, both clinical and commercial, are enormous. This is a young area of research and although progress has been rapid so far, there remain significant technical hurdles. In our view, there are two major issues that need to be resolved for there to be significant progress in this field.

First, it will be important to determine the most appropriate starting material from which to generate β-cells *in vitro*. Opinion is currently divided between embryonic and tissue stem cells as potential therapeutic progenitor cells, and both sources have inherent advantages and disadvantages. The greater proliferative capacity of ES cells makes them an attractive starting material, but their pluripotency may be a disadvantage because they tend to differentiate into a wide spectrum of cell types and current methods of ES cell differentiation have not yet produced the homologous populations of fully-differentiated β-cells required for transplantation therapy. Tissue stem cells show some degree of lineage-restriction, which, given an appropriate starting population, may make it easier to induce differentiation towards a β-cell phenotype. However, the limited replicative potential of tissue stem cell populations *in vitro* is a great disadvantage when the large numbers of cells required for replacement therapy of T1DM are taken into consideration.

The issue of optimising differentiation of stem cell populations towards a regulated insulin-secreting phenotype is also of great importance. Many current studies, particularly those using ES cells, attempt to drive stem cell differentiation towards β-cells using information gained from studies of pancreatic development: progress in this area clearly requires a more detailed understanding of the extracellular signals that regulate stem cell fate and the intracellular signalling pathways that govern the responses of cells to these extrinsic signals.

From the current perspective these issues are not insurmountable, and new technical approaches and achievements are continually being reported. If progress in stem cell research continues at its present pace the translation of these experimental studies into a clinical therapy for diabetes may occur in the not too distant future.

Acknowledgements

Our work on stem cells has been supported by Diabetes UK, Diabetes Research and Wellness Foundation, the Juvenile Diabetes Research Foundation, Guy's and St. Thomas' Charitable Foundation, the Diabetes Foundation and the EU Framework V programme.

References

Abe K, Niwa H, Iwase K, Takiguchi M, Mori M, Abe SI, Abe K, Yamamura KI (1996) Endoderm-specific gene expression in embryonic stem cells differentiated to embryoid bodies. *Exp. Cell Res.*, *229,* 27-34.

Abraham EJ, Leech CA, Lin JC, Zulewski H and Habener JF (2002) Insulinotropic hormone glucagon-like peptide-1 differentiation of human pancreatic islet derived progenitor cells into insulin-producing cells. *Endocrinology 1453*, 3152-3161.

Ahlgren U, Jonsson J, Edlund H. (1996) The morphogenesis of the pancreatic mesenchyme is uncoupled from that of the pancreatic epithelium in IPF1/PDX1-deficient mice. *Development. 122*, 1409-16.

Apelqvist A, Li H, Sommer L, Beatus P, Anderson DJ, Honjo T, Hrabe de Angelis M, Lendahl U, Edlund H (1999) Notch signalling controls pancreatic cell differentiation. *Nature. 400*, 877-81.

Asahara T, Masuda H, Takahashi T, Kalka C, Pastore C, Silver M, Kearne M, Magner M and Isner JM (1999) Bone marrow origin of endothelial progenitor cells responsible for postnatal vasculogenesis in physiological and pathological neovasularisation. *Circ. Res. 85*, 221-228.

Assady S, Maor G, Amit M, Itskovitz-Eldor J, Skorecki KL and Tzukerman M (2001) Insulin production by human embryonic stem cells. *Diabetes 50*, 1691-1697

Bai L, Meredith G and Tuch BE (2005) Glucagon-like peptide-1 enhances production of insulin in insulin-producing cells derived from mouse embryonic stem cells. *J. of Endocrinol. 186*, 343-352.

Barberi T, Klivenyi P, Calingasan NY, Lee H, Kawamata H, Loonam K, Perrier AL, Bruses J, Rubio ME, Topf N, tabar V, Harrison NL, Beal MF, Moore MAS and Struder L (2003) Neural subtype specification of fertilization and nuclear transfer embryonic stem cells and application in parkinsonian mice. *Nat Biotechnol. 21*, 1200-1207.

Blyszczuk P, Czyz J, Kania G, Wagner M, Roll U, St Onge L, Wobus AM (2003) Expression of Pax4 in embryonic stem cells promotes differentiation of nestin-positive progenitor and insulin-producing cells. *Proc. Natl. Acad. Sci. USA 100*, 998-1003.

Bonner-Weir S, Taneja M, Weir GC, Tatarkiewicz K, Song KH, Sharma A and O'Neil JJ (2000) In vitro cultivation of human islets from expanded ductal tissue. *Proc. Natl. Acad. Sci. USA 97*, 7999-8004.

Brolen GK, Heins N, Edsbagge J, Semb H. (2005) Signals From the Embryonic Mouse Pancreas Induce Differentiation of Human Embryonic Stem Cells Into Insulin-Producing β-Cell-Like Cells. *Diabetes. 54*, 2867-74.

Burns CJ, Minger SL, Hall S, Roderigo-Milne H, Ramracheya RD, Evans ND, Persaud SJ and Jones PM (2005) The in vitro differentiation of rat neural stem cells into an insulin-expressing phenotype. *Biochem. Biophys. Res. Comm. 326*, 570-577.

Burns CJ, Patel M, Minger SL, Pickering S, Braude PR, Kararigas G, Kitsou-Mylona I, Wu Y, Jones PM and Persaud SJ (2004) Human embryonic stem cells differentiate to express pancreatic and β-cell-specific genes. *Diabetic Medicine 21*, P2

Choi JB, Uchino H, Azuma K, Iwashita N, Tanaka Y, Mochizuki H, Migita M, Shimanda T, Kawamori R and Watada H (2003) Little evidence of transdifferentiation of bone marrow-derived cells into pancreatic beta cells. *Diabetologia 46*, 1366-1374.

Clarke DL, Johansson CB, Wilbertz J, Veress B, Nilsson E, Karlstrom H, Lendahl U and Frisen J (2000) Generalized potential of adult neural stem cells. *Science 288*, 1660-1663

Devasker SU, Sadiq HF, Holtzclaw L and George M (1993a) The developmental pattern of rabbit brain insulin and insulin-like growth factor receptor expression. *Brain Res. 605*, 101-109

Devasker SU, Singh BS, Carnaghi LR, Rajakumar PA and Giddings SJ (1993b) UInsulin II gene expression in rat central nervous system *Regul. Peptides 48*, 55-63.

Dor Y, Brown J, Martinez OI, Melton DA (2004) Adult pancreatic beta-cells are formed by self-duplication rather than stem cell differentiation. *Nature 429*, 41-46.

Ferber S, Halkin A, Cohen H, Ber I, Einav Y *et al.* (2000) Pancreatic and duodenal homeobox gene 1 induces expression of insulin genes in liver and ameliorates streptozotocin-induced hyperglycemia. *Nature Medicine 6*, 568-572.

Ferrari G, Cusella-De Angelis G, Coletta M, Paolucci E, Stornaiuolo A, Cossu G and Mavillo F (1998) Muscle regeneration by bone marrow-derived myogenic progenitors. *Science 279*, 1528-1530.

Gaglia JL, Shapiro JAM and Weir GC (2005) Islet transplantation: Progress and Challenge. *Arch of Med Res 36*, 273-280.

Gershengorn MC, Hardikar AA, Wei C, Geras-Raaka E, Marcus-Samuels B and Raaka BM (2004) Epithelial-to-Mesenchymal transition generates proliferative human islet precursor cells. *Science 306*, 2261-2264.

Hanley NA (2002) Beta-cell differentiation during human development does not rely on nestin-positive precursors: implications for stem cell-derived replacement therapy. *Diabetologia 45*, 1045-1047

Hay ED (1995) An overview of epithelio-mesenchymal transition. *Acta. Anat 154*, 8-20.

Horb M, Shen CN, Tosh D and Slack JMW (2003) Experimental conversion of pancreas to liver *Current Biology 13*, 105-115.

Hori Y, Rulifson I C, Tsai B, Heit JJ, Cahoy JD, Kim SK (2002) Growth inhibitors promote differentiation of insulin-producing tissue from embryonic stem cells. *Proc. Nat. Acad. Sci, USA*, *99*, 16105–16110.

Hori Y, Gu X, Xie X, Kim SK. (2005) Differentiation of insulin-producing cells from human neural progenitor cells. *PLoS Med. 2*, 103.

Houard N, Rousseau GG and Lemaigre FP (2003) HNF-6-independent differentiation of mouse embryonic stem cells into insulin-producing cells. *Diabetologia*, *46*, 378-385.

Humphrey RK, Bucay N, Beattie GM, Lopez AD, Messam CA, Cirulli V and Hayek A (2003) Characterization and isolation of promoter-defined nestin-positive cells from the human fetal pancreas. *Diabetes 52*, 2519-2525

Hunziker E and Stein M (2000) Nestin-expressing cells in the pancreatic islets of Langerhans. *Biochem. Biophys, Res. Comm. 271*, 116-119

Hwang WS, Roh SI, Lee BC, Kang SK, Kwon DK, Kim S, Kim SJ, Park SW, Kwon HS, Lee CK, Lee JB, Kim JM, Ahn C, Paek SH, Chang SS, Koo JJ, Yoon HS, Hwang JH, Hwang YY, Park YS, Oh SK, Kim HSPark JH, Moon SY and Schatten G (2005) Patient-specific embryonic stem cells derived from human SCNT blastocysts. *Science 308*, 1777-1783.

Ianus A, Holz GG, Theise ND and Hussain MA (2003) In vivo derivation of glucose-competent pancreatic endocrine cells from bone marrow without evidence of cell fusion. *J. Clin. Invest. 111*, 843-850

Jahr H and Bretzel B (2003) Insulin-positive cells in vitro generated from rat bone marrow stromal cells. *Transplantation Proc. 35*, 2140-2141

Jones PM & Persaud SJ (1998) Protein kinases, protein phosphorylation and the regulation of insulin secretion from pancreatic β-cells. *Endocrine Reviews 19*, 429-461

Kahan BW, Jacobson LM, Hullett DA, Ochoada JM, Oberley TD, Lang KM, Odorico JS. (2003) Pancreatic precursors and differentiated islet cell types from murine embryonic stem cells: an in vitro model to study islet differentiation. *Diabetes 52*, 2016-24.

Kania G, Blyszczuk P, Wobus AM. (2004) The generation of insulin-producing cells from embryonic stem cells--a discussion of controversial findings. *Int J Dev Biol. 48,* 1061-4.

Kim D, Gu Y, Ishii M, Fujimiya M, Qi M, Nakamura N, Yoshikawa T, Sumi S, Inoue K. (2003) In vivo functioning and transplantable mature pancreatic islet-like cell clusters differentiated from embryonic stem cell. *Pancreas. 27*, e34-41.

Klein T, Ling Z, Heimburg H, Madsen OD, Heller RS and Serup P (2003) Nestin is expressed in vascular endothelial cells in the adult human pancreas,.*J. Histochem. Cytochem. 51*, 697-706

Kojima H, Fujimiya M, Matsumura K, Younan P, Imaeda H, Maeda M and Chan L (2003) NeuroD-betacellulin gene therapy induces islet neogenesis in the liver and reverses diabetes in mice. *Nature Medicine 9*, 596-603.

Kopen GC, Prockop DJ and Phinney DG (1999) Marrow stromal cells migrate throughout forebrain and cerebellum, and they differentiate into astrocytes after injection into neonatal mouse brains. *Proc. Natl. Acad. Sci USA 96*, 10711-10716.

Ku HT, Zhang N, Kubo A, O'Connor R, Mao M, Keller G and Bromberg JS. (2004) Committing embryonic stem cells to early endocrine pancreas in vitro. *Stem Cells. 22*, 1205-17.

Lechner A and Habner JF (2003) Stem/progenitor cells derived from adult tissues: potential for the treatment of diabetes mellitus. *Am. J. Physiol. Endocrinol. Metab. 284*, E259-E266

Lechner A, yang YG, Blacken RA, Wang l, Nolan AL and Habener JF (2004) No evidence for significant transdifferentiation of bone marrow into pancreatic β-cells in vivo. *Diabetes 53*, 616-623.

Leon-Quinto T, Jones J, Skoudy A, Burcin M and Soria B (2004) In vitro directed differentiation of mouse embryonic stem cells into insulin-producing cells. *Diabetologia 47,* 1442-1451.

Li W-C, Horb ME, Tosh D and Slack JMW (2005) In vitro transdifferentiation of hepatoma cells into functional pancreatic cells. *Mech of Dev. 122*, 835-847.

Lumelsky N, Blondel O, Laeng P, Velasco I, Ravin R, McKay R. (2001) Differentiation of embryonic stem cells to insulin-secreting structures similar to pancreatic islets. *Science 292*, 1389-1394.

McGrath KE and Palis J (1997) Expression of homeobox genes, including an insulin promoting factor, in the murine yolk sac at the time of hematopoietic initiation. *Mol. Reprod. Dev. 48*, 145-153

Milne HM, Burns CJ, Kitsou-Mylona I, Luther MJ, Minger SL, Persaud SJ and Jones PM (2005) Generation of insulin-expressing cells from mouse embryonic stem cells. *Biochem Biophys Res Commun. 328*, 399-403.

Minger SL, Fisher LJ, Ray J and Gage FH (1996) Long-term survival of transplanted basal forebrain cells following in vitro propagation with fibroblast growth factor-2. *Exp. Neurol. 141*, 12-24

Micallef SJ, Janes ME, Knezevic K, Davis RP, Elefanty AG and Stanley EG (2005) Retinoic acid induces Pdx-1 positive endoderm in differentiating mouse embryonic stem cells. *Diabetes 54*, 301-305.

Miyazaki S, Yamato E and Miyazaki J-I (2004) Regulated expression of pdx-1 promotes in vitro differentiation of insulin producing cells from embryonic stem cells. *Diabetes 53*, 1030-1037

Moriscot C, De Fraipont F, Richard MJ, Marchand M, Savatier P, Bosco D, Favrot M and Benhamou PY (2005) Human bone marrow mesenchymal stem cells can express insulin and key transcription factors of the endocrine pancreas developmental pathwat upon genetic and/or microenvironmental manipulation in vitro. *Stem Cells 23*, 594-604.

Moritoh, Y., Yamato, E., Yasui, Y., Miyazaki, S. and Miyazaki, J.I. (2003) Analysis of insulin-producing cells during in vitro differentiation from feeder-free embryonic stem cells. *Diabetes 52*, 1163-1168

Murray P and Edgar D (2001)The regulation of embryonic stem cell differentiation by leukaemia inhibitory factor (LIF). *Differentiation 68*, 227-234.

Odom DT, Zizlsperger N, Gordon DB, Bell GW, Rinaldi HJ, Murray HL, Volkert TL, Schreiber J, Rolfe PA, Gifford DK, Fraenkel E, Bell GI and Young RA (2004) Control of pancreas and liver gene expression by HNF transcription factors *Science 303*, 1378-1381.

Paek HJ, Morgan JR, Lysaght MJ. (2005) Sequestration and synthesis: the source of insulin in cell clusters differentiated from murine embryonic stem cells. *Stem Cells. 23*, 862-7.

Pereira RF, Halford KW, O'Hara MD, Leeper DB, Sokolov BP, Pollard MD, Bagasra O and Prockop DJ (1995) Cultured adherent cells from marrow can serve as long-lasting precursor cells for bone, cartilage, and lung in irradiated mice. *Proc. Natl. Acad. Sci. USA 92*, 4857-4861.

Rajagopal J, Anderson WJ, Kume S, Martinez OI, Melton DA. (2003) Insulin staining of ES cell progeny from insulin uptake. *Science 299*, 363.

Ramiya VK, Maraist M, Arfors KE, Schatz DA, Peck AB, Cornelius JG (2000) Reversal of insulin-dependent diabetes using islets generated in vitro from pancreatic stem cells. *Nature Medicine 6*, 278-282

Rau K, Muglia L and Locker J (1989) Insulin-gene expression in extrafetal membranes of rats. *Diabetes 38*, 39-43

Rossant J. (2004) Lineage development and polar asymmetries in the peri-implantation mouse blastocyst. *Semin Cell Dev Biol. 15*, 573-81.

Ryan EA, Lakey JRT, Rajotte RV, Korbutt GS *et al.* (2001) Clinical outcomes and insulin secretion after islet transplantation with the Edmonton Protocol. Diabetes *50*, 710-719

Sang U H, Alilain W and Saljooque F (2002) Fetal brain progenitor cells transdifferentiate to fates outside the nervous system. *Mol. Endocrinol. 16(11)*, 2645-2656

Sapir T, Shternhall K, Meivar-Levy I *et al.*, (2005) Cell-replacement therapy for diabetes: Generating functional insulin-producing tissue from adult human liver cells. *Proc. Natl. Acad. Sci. USA. 102*, 7964-9.

Scharfmann R (2000) Control of early development of the pancreas in rodents and humans: implications of signals from the mesenchyme. *Diabetologia 43*, 1083-1092

Schwartz RE, Reyes M, Koodie L, Jiang Y, Blackstad M, Lund TY, Lenvik T, Johnson S, Hu WS and Verfaille CM (2002) *J. Clin Invest. 109*, 1291-1302.

Seaberg RM, Smukler SR, Kieffer TJ *et al.,* (2004) Clonal identification of multipotent precursors from adult mouse pancreas that generate neural and pancreatic lineages. *Nature Biotechnology 22* 1115-1124.

Segev H, Fishman B, Ziskind A and Itskovitz-Eldor J (2004) Differentiation of human embryonic stem cells into insulin producing clusters. *Stem Cells 22*, 265-274

Selander L and Edlund H (2002) Nestin is expressed in mesenchymal and not epithelial cells of the developing mouse pancreas. *Mechanisms of Development 113*, 189-19

Shapiro AMJ, Lakey JRT, Ryan EA, Korbutt GS *et al.* (2000) Islet transplantation in seven patients with Type 1 diabetes mellitus using a glucocorticoid-free immunosuppressive regime. *N. Engl. J. Med 343*, 230-238.

Shi Y, Hou L, tang F, Jiang W, Wang P, Ding M and Deng H (2005) Inducing embryonic stem cells to differentiate into pancreatic β-cells by a novel three-step approach with activin A and all-trans retinoic acid. *Stem Cells 23*, 656-662.

Shiroi A, Yoshikawa H, Fukui H, Ishizaka S *et al.* (2002) Identification of insulin-producing cells derived from embryonic stem cells by zinc-chelating dithizone. *Stem Cells 20,* 284-292.

Shiroi A, Ueda S, Ouji Y, Saito K, Moriya K, Sugie Y, Fukui H, Ishizaka S, Yoshikawa M. (2005) Differentiation of embryonic stem cells into insulin-producing cells promoted by Nkx2.2 gene transfer. *World J Gastroenterol. 11*, 161-6.

Sipione S, Eshpeter A, Lyon JG, Korbutt GS and Bleackley RC (2004) Insulin expressing cells from differentiated embryonic stem cells are not beta cells. *Diabetologia 47*, 499-508

Soria B, Roche E, Berná G, León-Quinto T, Reig JA and Martín F (2000) Insulin-secreting cells derived from embryonic stem cells normalize glycemia in streptozotocin-induced diabetic mice. *Diabetes 49*, 157-162

Street CN, Lakey JRT, Seeberger K, Helms L, Rajotte RV, Sahpiro AMJ and Korbutt GS (2004) Heterogenous expression of nestin in human pancreatic tissue precludes its use as an islet precursor marker. *J. Endocrinol. 180*, 213-225

Suzuki A, Zheng Y-W, Kaneko S, Onodera M, Fukao K, Nakauchi H and Taniguchi H (2002) Clonal identification and characterization of self-renewing pluripotent stem cells in the developing liver. *J. Cell. Biol. 156*, 173-184.

Tang DQ, Cao LZ, Burkhardt BR, Xia CQ, Litherland SA, Atkinson MA and Yang LJ (2004) In vivo and In vitro characterisation of insulin-producing cells obtained from murine bone marrow. *Diabetes 53*, 1721-1732.

Thiery JP (2002) Epithelial-mesenchymal transitions in tumour progression. *Nat. Rev. Cancer* 2, 442-454.

Thiery JP (2003) Epithelial-mesenchymal transitions in development and pathologies. *Curr. Opin. Cell Biol. 15*, 740-746.

van der Laan LJW, Lockey C, Griffeth BC, Frsier FS, Wilson CA, Onions DE, Hering BJ, Long Z, Otto E, Torbett BE and Salomon DR (2000) Infection by porcine endogenous retrovirus after islet xenotransplantation in SCID mice. *Nature (London) 407*, 90-94

Wagers AJ, Sherwood RI, Christensen JL and Weissman IL (2002) Little evidence for developmental plasticity of adult hematopoietic stem cells. *Science 297*, 2256-2259.

Wilson ME, Scheel D and German MS. (2003) Gene expression cascades in pancreatic development *Mech. Dev. 120*, 65-80

Yang L, Li S, Hatch H, Ahrens A, Cornelius JG, Petersen BE and Peck AB (2002) In vitro trans-differentiation of adult hepatic stem cells into pancreatic endocrine hormone-producing cells. *PNAS 99*, 8078-8083

Yang X, Kow L-M, Funabashi T and Mobbs CV (1999) Hypothalamic glucose sensor. *Diabetes 48*, 1763-1772

Yoshida S, Ishikawa F, Kawano N, Shimoda K, Nagafuchi S, Shimoda S, Yasukawa M, Kanemaru T, Ishibashi H, Shultz LD, Harada M. (2005) Human cord blood-derived cells generate insulin-producing cells in vivo. *Stem Cells. 23*, 1409-16.

Zalzman, M, Gupta s, Giri R, Berkovich I, Sappal BS, Karnieli O, Zern MA, Fleischer N and Efrat S (2003) Reversal of hyperglycaemia in mice by using human expandable insulin-producing cells differentiated from fetal liver progenitor cells. *PNAS 100*, 7253-7258

Zulewski H, Abraham EJ, Gerlach MJ, Daniel PB, Moritz W, Muller B, Vallejo M, Thomas MK and Habener JF (2001) Multipotential nestin-positive stem cells isolated from adult pancreatic islets differentiate ex vivo into pancreatic endocrine, exocrine, and hepatic phenotypes. *Diabetes 50*, 521-533

In: Frontiers in Stem Cell Research
Editor: Julia M. Spanning pp. 173-185

ISBN 1-60021-294-8

Chapter VII

CELL THERAPY WITH BONE MARROW-DERIVED STEM CELLS FOR CARDIAC REPAIR OR REGENERATION

Young-sup Yoon*

Division of Cardiovascular Research, Caritas St. Elizabeth's Medical Center, Tufts University School of Medicine, Boston, Massachusetts, USA

ABSTRACT

Since the first experiments of cell transplantation into heart were performed in the early 1990s, the identification of adult stem cells has triggered attempts to regenerate damaged heart tissue by cellular transplantation. Until recently, a multitude of adult stem or progenitor cells from various tissues have been proposed to meet this end. Bone marrow, especially, has emerged as the most promising source for stem and progenitor cells because bone marrow, besides the organ of hematopoietic maintenance, contains a complex assortment of stem and progenitor cells. A large body of provocative experimental evidence for vascular and myocardial regeneration by these cells has furthered enthusiasm for their use, yet many questions remain unanswered in this new field of research regarding the therapeutic potential and the mechanisms responsible for the observed therapeutic effects. In this review article, we will discuss the therapeutic capacity of currently available representative bone marrow-derived stem and progenitor cells for treating ischemic heart diseases.

Keywords: Bone Marrow, Stem Cells, Regeneration, Ischemia, Infarction, Angiogenesis, Vasculogenesis, Myogenesis, Adult

* Correspondence concerning this article should be addressed to Young-sup Yoon, MD, PhD Division of Cardiovascular Research, Caritas St. Elizabeth's Medical Center, Tufts University School of Medicine 736 Cambridge Street, Boston, Massachusetts 02135, USA. Phone: (617) 562-7523; Fax: (617) 562-7506; E-mail:young.yoon@tufts.edu.

Despite many breakthroughs in cardiovascular medicine, ischemic heart diseases such as myocardial infarction (MI) and congestive heart failure (CHF) remain among the most prominent health challenges worldwide [1]. In the United States, for example, one of the major contributors to the development of CHF is MI, which occurs in nearly 1.1 million Americans each year and CHF afflicts 4.8 million people, with 550,000 new cases a year. Moreover, in spite of the medical and surgical treatments, prognosis is poor with a median survival after onset of only 1.7 years in men and 3.2 years in women [2,3]

The longstanding axiom has been that the myocardium has a limited capacity for self regeneration [4,5], and the irreversible loss of cardiac muscle and accompanying scar formation following diseases such as MI leads to progressive ventricular remodeling and CHF [6]. Currently no medication or procedure used clinically has shown efficacy in replacing myocardial scar with functioning contractile tissue. Therefore, given the major morbidity and mortality associated with CHF, new approaches have been sought to address the principal pathophysiologic deficits responsible for its development, i.e. loss of vessels and cardiomyocytes [7]. Recently, the identification of adult stem or progenitor cells capable of contributing to tissue regeneration has raised the possibility that cell therapy could be employed for repair of damaged myocardium. Several cell sources including bone marrow (BM) have been explored in an effort to regenerate myocardium [8-10].

Recent investigations suggested bone marrow cells as a potential source for adult stem or progenitor cells. BM-derived stem cells appear to have the capacity to repopulate many nonhematopoietic tissues, such as neuroectodermal cells (those from which the nervous system are derived) [11-13], skeletal myoblasts [14-16], cardiomyocyte [17,18], endothelium [17-19], and lung, gut and skin epithelia [20]. This plasticity of BM-derived cells is further supported by a series of reports on human BM transplanted tissues, which has shown that cells of donor origin can be detected in the extra-hematopoietic tissue of recipients including brain [21] and heart [22-25] as fully differentiated cells incorporated into host tissue. Furthermore, simplified approaches using unselected [26] or selected BM or peripheral blood mononuclear cell fractions [27,28], based on the premise that BM or peripheral blood cells would include stem and progenitor cell populations which could augment neovascularization in ischemic tissue, were reported to be successful in improving cardiac function following various models of myocardial ischemia. The promising results from initial experimental studies promoted the initiation of clinical trials [29-33] although the extent of multilineage potential has been challenged [34-36]. Various strategies were adopted by different groups according to the cell, disease states and route of delivery [29-32,37-40]. The question of whether the infused cells in the heart preferentially act via improving tissue perfusion or by regenerating cardiomyocytes as well remains unresolved.

In this review, we would like to discuss the therapeutic potential of the following representative bone marrow-derived stem and progenitor cells in cardiovascular diseases; hematopoietic stem cells (HSCs), endothelial progenitor cells (EPCs), mesenchymal stem cells (MSCs), multipotent adult progenitor cells (MAPCs), and human bone marrow-derived multipotent stem cells (hBMSCs).

Hematopoietic Stem Cells

Besides hematopoietic potential, recent studies suggested the possibility that specific SC populations derived from murine BM could regenerate most of the key elements of myocardium and ameliorate ischemic cardiac dysfunction. The more versatile potential of BM-derived HSCs has been documented with the use of side population cells (cells defined by their ability to expel Hoechst dye) [41] and with lineage-negative, *c-kit* positive (cells that express a specific protein known as a tyrosine kinase which adds phosphate groups to the amino acid tyrosine in other proteins; Lin^- *c-kit* $^+$) cells [42]; both of which differentiated into ECs, SMCs and cardiomyocytes in a murine MI model [17,43]. However, other studies have claimed that HSCs do not readily acquire a cardiac phenotype in injured heart, despite the improvement in ventricular function in the HSCs transplanted group [34,35]. These studies were all performed with murine HSC and currently no studies have been performed with the human counterpart of these SCs. In addition, these HSCs consisted of heterogeneous population of SCs. Therefore, it is uncertain whether single HSCs can participate in angio-vasculogenesis and/or cardiomyogenesis. In terms of clinical application, the scarcity of these HSCs in BM and the unavailability of culture-expansion techniques together with controversial reports may limit the therapeutic application of HSCs in ischemic heart diseases.

Endothelial Progenitor Cells

Angiogenesis and vasculogenesis are responsible for the development of the vascular system in the development [44]. Vasculogenesis refers to the de novo development of blood vessels from angioblasts or vascular stem cells [45,46]. In contrast, angiogenesis is the formation of new capillaries from preexisting blood vessels through migration and proliferation of endothelial cells, and sprouting and branching of capillaries [47,48]. Until late 1990s, vasculogenesis had been thought to be restricted to embryonic developmental stages [49]. However, this notion was dismantled by the discovery of existence of endothelial progenitor cells (EPCs) in post-natal BM [50,51]. It is now known that BM-derived EPCs circulate in peripheral blood, home to and incorporate into foci of neovascularization [52], and increase in numbers in response to tissue ischemia in adult mammals [53].

EPCs subsequently were isolated from human umbilical cord blood, BM-derived mononuclear cells, and $CD34^+$ or $CD133^+$ hematopoietic SCs [19,51,54,55]. These cells differentiate into endothelial cells, as shown by expression of various endothelial proteins (von Willebrand factor, endothelial nitric oxidase synthase, VE-cadherin), uptake of DiI-acetylated low-density lipoprotein and binding of lectin. The systemically administered ex-vivo expanded EPCs following MI incorporate into foci of myocardial neovascularization and have a favorable impact on preservation of left ventricular function [56]. Direct intramyocardial injection of human CD34+ cells has also been shown to induce new blood vessel formation in the infarct vascular bed and proliferation of preexisting vasculature after experimental MI of rats [57,58]. However, despite extensive studies of EPCs in both experimental and clinical field, the origin and heterogeneity of their nature are still controversial.

Mesenchymal Stem Cells

Another source of adult SCs being explored for ischemic cardiovascular disease is nonhematopoietic MSCs (or marrow stromal cells). MSCs can be derived from adult BM and have multi-lineage differentiation capacity [59,60]. Both in vivo and in vitro studies have demonstrated that MSCs can undergo cardiomyogenic differentiation [61,62]. In animal models of MI, it has also been documented that MSCs can undergo milieu-dependent [62-64] and time-dependent [65] differentiation and express cardiomyogenic phenotypes. This cardiomyogenic differentiation capacity was demonstrated with animal as well as human MSCs. However, the therapeutic significance of vasculo-myogenic differentiation has been questioned due to the poor cell viability following MSC transplantation [62-65]. Therefore, other groups attempted gene transfection into MSCs with the use of heme oxygenase-1 plasmid [66] or adenoviral Akt [67]. These studies showed that improved cell survival by preventing apoptosis enhanced cardiac fucntion. Intriguingly, in a murine MI model, it was reported that culture-conditioned media of MSCs exert similar therapeutic effects as MSCs, suggesting that paracrine action could be a major therapeutic mechanism [68]. Also another group demonstrated that in a model of murine hindlimb ischemia, MSC transplantation enhanced tissue repair via secretion of multiple cytokines including VEGF, bFGF and PlGF rather than incorporation of MSCs into new or remodeling tissues [69-72].

The advantage of MSCs for potential clinical application lies in the ease of acquisition, isolation and high expansion capability in culture. This extensive capacity for clinical scale expansion in vitro has facilitated the development of preclinical models as well as clinical studies designed to assess the safety, feasibility and efficacy of MSC transplantation in a variety of diseases. However, the use of animal serum for culture-expansion limits the growth of pilot clinical trials using MSCs in humans.

Multipotent Adult Progenitor Cells

Recently, a subset of BM-derived SCs referred to as multipotent adult progenitor cells (MAPCs) has been shown to possess the most versatile potential of adult SCs to date. MAPCs could be clonally expanded, proliferate indefinitely without obvious genetic instability and differentiate into cells of all three embryonic germ layers in vitro [73-76]. When injected in the blastocyst, chimerism could be detected in most if not all somatic tissues. On transplantation into a non-irradiated or minimally irradiated host, MAPCs could engraft and differentiate into the hematopoietic lineage as well as other lineages. However, MAPCs showed minimal engraftment into the myocardium after being injected into the embryonic blastocyst. In addition MAPCs have not been tested for therapeutic or differentiation potential in animal models of cardiovascular diseases. Furthermore, the requirement of multiple expensive cytokines to culture MAPCs renders them less attractive for clinical use.

Bone Marrow-Derived Multipotent Stem Cells

Another stem cell population, BMSCs was clonally isolated from human and rat beginning at the single-cell level from mixed total BM cell culture [77]. Clonally isolated BMSCs exhibited unlimited self-renewal and differentiated into cells of triple-lineage. None of the characteristic marker panels (e.g. CD34, CD44, CD90, CD105 and CD117) defining HSCs, MSCs, and MAPCs matches the profile of human BMSCs. The minimal expression of surface molecules appears to be a prerequisite for the plasticity of hBMSC as shown in other adult multipotent SCs. In vivo experiment, the transplantation of hBMSCs into acutely infarcted myocardium demonstrated the attenuation of cardiac dysfunction both by de no differentiation into heart tissue and by paracrine, i.e. non-cellular differentiation mechanisms. Also engrafted hBMSCs have been demonstrated to provoke host myocardial cell proliferation such as ECs, SMCs and CMCs suggesting endogenous angio/arteriogenesis and cardiomyogenesis could occur following SCs transplantation [77,78].

The potential advantages of hBMSCs with respect to their potential for treating cardiovascular diseases include their culture-expandability, the higher engraftment rate in ischemic and infarcted myocardium, and the demonstrated multi-functional capacity for cardiovascular tissue regeneration. On the other hand, the difficulty of culturing techniques and the use of large amount of animal serum are some of the hurdles to be overcome for clinical use.

Table 1. Characteristics of various bone marrow-derived stem cells

	EPC	HSC			MSC	MAPC	BMSC
		CD34/CD133	SP	Lin^{-}*c-kit*$^{+}$			
Single cell derived clonal origin	-	-	-	-	-	+	+
Vasculogenesis	+	+	+	+	±	+	+
Cardiomyogenesis	±	-	+	+	+	?	+
Paracrine effects	+	?	?	?	+	?	+

EPC, endothelial progenitor cells; HSC, hematopoietic stem cells; CD34, CD34 positive cells; CD133, CD133 positive cells; SP, side population cells; Lin^{-}*c-kit*$^{+}$, lineage negative *c-kit* positive cells; MSC, mesenchymal stem cells; MAPC, multipotent adult progenitor cells; BMSC, bone marrow-derived multipotent stem cells.

Mechanisms Mediating Therapeutic Effects of BM-Derived Stem or Progenitor Cells

Although stem cell therapy has emerged as a promising therapeutic modality for treating ischemic heart diseases, there has been a controversy on the underlying mechanisms. One of the most intense debate lies in the plasticity of BM-derived stem cells, i.e., whether transdifferentiation truly occurs or not [79,80] The plasticity denotes the capability of SCs to differentiate into the cell types other than their cell origin. While initial studies demonstrated that certain population of BM-derived stem cells into cardiomyocytes, ECs and SMCs, other investigators claimed that transdifferentiation can hardly or never occur [34,35], and the other investigators argued that transdifferentiation may occur due to fusion rather than true transdifferentiation [81,82] Conventionally, it was accepted that transdifferentiation of SCs occur when transplanted SCs labeled with genetic markers (GFP or LacZ) or chemical dyes express (was colocalized with) specific marker proteins of the cell of interest. Initial studies claimed transdifferentiation of BM-derived SCs into myocardial cells by using these criteria [17,18,83]. Later studies, however argued that fusion between stem and other cells may also appear as transdifferentiated cells [81,82]. Through an approach of labeling both stem and recipient cells, investigators reported that a majority or all of the apparent transdifferentiation develop due to fusion of stem and recipient cells [84,85]. In fact, recent studies have suggested that both fusion and transdifferentiation are responsible for the phenotypic changes of stem cells [77]. Furthermore, a paracrine mechanism was emerged as an alternative or additional mechanism to explain therapeutic effects because recent studies indicated that whether transdifferentiation truly occured or not, the extent of myogenesis and/or vasculogenesis from transplanted BM-derived SCs was not suffient to explain the beneficial effects of cell transplantation [71,77,86] The proposed paracrine hypothesis is that BM-derived SCs secrete multiple cytokines as a paracrine manner to promote angiogenesis [71,86,87] and cardiomyogenesis [77], protect myocardial apoptosis [68,77], and induce endogenous resident SC migration and proliferation [88]. At this moment, it may not be easy to dissect the relative contribution of each component due to the difference of animal models used, type of cells used, route of delivery and so on. Taken together, multiple mechanisms listed above appear to be mediating therapeutic effects of BM-derived stem cells (Figure 1).

FUTURE PERSPECTIVE

BM-derived SC research has been making rapid progress and pilot clinical trials have been already started. However, there are many remaining questions with regard to their therapeutic use. First, the specific role of SCs in the different disease states may be misunderstood or overestimated. The type of cardiovascular diseases and the components requiring regeneration should be foremost considered. Second, the development of therapeutic strategies to induce both neovascularization and cardiomyogenesis are required since both processes are intimately related in successful myocardial remodeling and regeneration. Third, we need to acknowledge that not all the BM-derived SCs possess the same therapeutic capacity and efficacy. Further investigations are needed to compare the therapeutic effects and the underlying mechanisms of different types of stem and progenitor cells, and the advantage of mixed cell therapy may need to be investigated. Fourth, the dose, route of delivery and the timing of therapy needs to be determined by more sophisticated approaches. Then, the problem of extensive death after cell transplantation should be addressed for improving therapeutic efficacy. Finally, the duration of benefits following SC transplantation also has to be determined. In spite of these questions and hurdles, many investigators agree that stem or progenitor cell therapy will become an effective therapeutic option for treating various cardiac diseases.

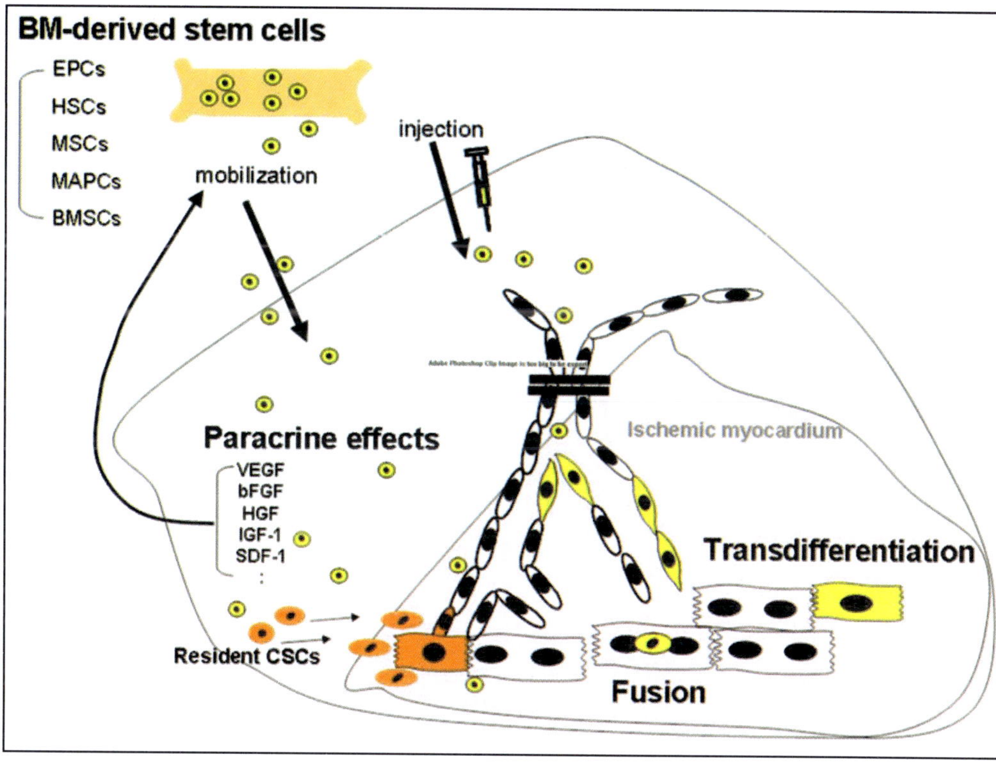

Figure 1. Schematic diagram of underlying mechanisms on therapeutic effects of bone marrow-derived stem cells. EPCs, endothelial progenitor cells; HSCs, hematopoietic stem cells which include CD34 and/or CD133 positive cells; CD133, side population cells, and lineage negative *c-kit* positive cells; MSCs, mesenchymal stem cells; MAPCs, multipotent adult progenitor cells; BMSCs, bone marrow-derived multipotent stem cells; CSCs, cardiac stem cells.

REFERENCES

[1] WHO. World Health Report. *www.who.int/whr/.*

[2] Association, A.H. 2002. *Heart and Stroke Statistical Update.* Dallas, TX: American Heart Association.

[3] Miller, L.W., and Missov, E.D. 2001. Epidemiology of heart failure. *Cardiol Clin 19*:547-555.

[4] MacLellan, W.R., and Schneider, M.D. 2000. Genetic dissection of cardiac growth control pathways. *Annu Rev Physiol 62*:289-319.

[5] Chien, K.R., and Olson, E.N. 2002. Converging pathways and principles in heart development and disease: CV@CSH. *Cell 110*:153-162.

[6] Pfeffer, M.A., and Braunwald, E. 1990. Ventricular remodeling after myocardial infarction. Experimental observations and clinical implications. *Circulation 81*:1161-1172.

[7] von Harsdorf, R., Poole-Wilson, P.A., and Dietz, R. 2004. Regenerative capacity of the myocardium: implications for treatment of heart failure. *Lancet 363*:1306-1313.

[8] Dimmeler, S., Zeiher, A.M., and Schneider, M.D. 2005. Unchain my heart: the scientific foundations of cardiac repair. *J Clin Invest 115*:572-583.

[9] Laflamme, M.A., and Murry, C.E. 2005. Regenerating the heart. *Nat Biotechnol 23*:845-856.

[10] Wollert, K.C., and Drexler, H. 2005. Clinical applications of stem cells for the heart. *Circ Res 96*:151-163.

[11] Brazelton, T.R., Rossi, F.M., Keshet, G.I., and Blau, H.M. 2000. From marrow to brain: expression of neuronal phenotypes in adult mice. *Science 290*:1775-1779.

[12] Mezey, E., Chandross, K.J., Harta, G., Maki, R.A., and McKercher, S.R. 2000. Turning blood into brain: cells bearing neuronal antigens generated in vivo from bone marrow. *Science 290*:1779-1782.

[13] Kopen, G.C., Prockop, D.J., and Phinney, D.G. 1999. Marrow stromal cells migrate throughout forebrain and cerebellum, and they differentiate into astrocytes after injection into neonatal mouse brains. *Proc Natl Acad Sci U S A 96*:10711-10716.

[14] Ferrari, G., Angelis, G.C.D., Coletta, M., Paolucci, E., Stornaiuolo, A., Gossu, G., and Mavilio, F. 1998. Muscle regeneration by bone marrow-derived myogenic progenitors. *Science 279*:1528-1530.

[15] Gussoni, E., Soneoka, Y., Strickland, C.D., Buzney, E.A., Khan, M.K., Flint, A.F., Kunkel, L.M., and Mulligan, R.C. 1999. Dystrophin expression in the mdx mouse restored by stem cell transplantation. *Nature 401*:390-394.

[16] LaBarge, M.A., and Blau, H.M. 2002. Biological progression from adult bone marrow to mononucleate muscle stem cell to multinucleate muscle fiber in response to injury. *Cell 111*:589-601.

[17] Orlic, D., Kajstura, J., Chimenti, S., Jakoniuk, I., Anderson, S.M., Li, B., Pickel, J., Mckay, R., Nadal-Ginard, B., Bodine, D.M., et al. 2001. Bone marrow cells regenerate infarcted myocardium. *Nature 401*:701-705.

[18] Jackson, K.A., Majka, S.M., Wang, H., Pocius, J., Hartley, C.J., Majesky, M.W., Entman, M.L., Michael, L.H., Hirschi, K.K., and Goodell, M.A. 2001. Regeneration of ischemic cardiac muscle and vascular endothelium by adult stem cells. *J Clin Invest 107*:1395-1402.

[19] Asahara, T., Masuda, H., Takahashi, T., Kalka, C., Pastore, C., Silver, M., Kearne, M., Magner, M., and Isner, J.M. 1999. Bone marrow origin of endothelial progenitor cells responsible for postnatal vasculogenesis in physiological and pathological neovascularization. *Circ Res 85*:221-228.

[20] Krause, D.S., Theise, N.D., Collector, M.I., Henegariu, O., Hwang, S., Gardner, R., Neutzel, S., and Sharkis, S.J. 2001. Multi-organ, multi-lineage engraftment by a single bone marrow-derived stem cell. *Cell 105*:369-377.

[21] van Praag, H., Schinder, A.F., Christie, B.R., Toni, N., Palmer, T.D., and Gage, F.H. 2002. Functional neurogenesis in the adult hippocampus. *Nature 415*:1030-1034.

[22] Quaini, F., Urbanek, K., Beltrami, A.P., Finato, N., Beltrami, C.A., Nadal-Ginard, B., Kajstura, J., Leri, A., and Anversa, P. 2002. Chimerism of the transplanted heart. *N Engl J Med 346*:5-15.

[23] Beltrami, A.P., Urbanek, K., Kajstura, J., Yan, S.M., Finato, N., Bussani, R., Nadal-Ginard, B., Silvestri, F., Leri, A., Beltrami, C.A., et al. 2001. Evidence that human cardiac myocytes divide after myocardial infarction. *N Engl J Med 344*:1750-1757.

[24] Laflamme, M.A., Myerson, D., Saffitz, J.E., and Murry, C.E. 2002. Evidence for cardiomyocyte repopulation by extracardiac progenitors in transplanted human hearts. *Circ Res 90*:634-640.

[25] Minami, E., Laflamme, M.A., Saffitz, J.E., and Murry, C.E. 2005. Extracardiac progenitor cells repopulate most major cell types in the transplanted human heart. *Circulation 112*:2951-2958.

[26] Fuchs, S., Baffour, R., Zhou, Y.F., Shou, M., Pierre, A., Tio, F.O., Weissman, N.J., Leon, M.B., Epstein, S.E., and Kornowski, R. 2001. Transendocardial delivery of autologous bone marrow enhances collateral perfusion and regional function in pigs with chronic experimental myocardial ischemia. *J Am Coll Cardiol 37*:1726-1732.

[27] Kamihata, H., Matsubara, H., Nishiue, T., Fujiyama, S., Tsutsumi, Y., Ozono, R., Masaki, H., Mori, Y., Iba, O., Tateishi, E., et al. 2001. Implantation of bone marrow mononuclear cells into ischemic myocardium enhances collateral perfusion and regional function via side supply of angioblasts, angiogenic ligands, and cytokines. *Circulation 104*:1046-1052.

[28] Shintani, S., Murohara, T., Ueno, T., Ikeda, H., Duan, J., and Imaizumi, T. 2000. Local transplantation of autologous bone marrow-derived mononuclear cells augments collateral vessel formation in ischemic hindlimb in rabbits. *Circulation 100*:I-406.

[29] Assmus, B., Schachinger, V., Teupe, C., Britten, M., Lehmann, R., Dobert, N., Grunwald, F., Aicher, A., Urbich, C., Martin, H., et al. 2002. Transplantation of Progenitor Cells and Regeneration Enhancement in Acute Myocardial Infarction (TOPCARE-AMI). *Circulation 106*:3009-3017.

[30] Strauer, B.E., Brehm, M., Zeus, T., Kostering, M., Hernandez, A., Sorg, R.V., Kogler, G., and Wernet, P. 2002. Repair of infarcted myocardium by autologous intracoronary mononuclear bone marrow cell transplantation in humans. *Circulation 106*:1913-1918.

[31] Kang, H.J., Kim, H.S., Zhang, S.Y., Park, K.W., Cho, H.J., Koo, B.K., Kim, Y.J., Soo Lee, D., Sohn, D.W., Han, K.S., et al. 2004. Effects of intracoronary infusion of peripheral blood stem-cells mobilised with granulocyte-colony stimulating factor on left ventricular systolic function and restenosis after coronary stenting in myocardial infarction: the MAGIC cell randomised clinical trial. *Lancet 363*:751-756.

[32] Wollert, K.C., Meyer, G.P., Lotz, J., Ringes-Lichtenberg, S., Lippolt, P., Breidenbach, C., Fichtner, S., Korte, T., Hornig, B., Messinger, D., et al. 2004. Intracoronary autologous bone-marrow cell transfer after myocardial infarction: the BOOST randomised controlled clinical trial. *Lancet 364*:141-148.

[33] Patel, A.N., Geffner, L., Vina, R.F., Saslavsky, J., Urschel, H.C., Jr., Kormos, R., and Benetti, F. 2005. Surgical treatment for congestive heart failure with autologous adult stem cell transplantation: a prospective randomized study. *J Thorac Cardiovasc Surg 130*:1631-1638.

[34] Balsam, L.B., Wagers, A.J., Christensen, J.L., Kofidis, T., Weissman, I.L., and Robbins, R.C. 2004. Haematopoietic stem cells adopt mature haematopoietic fates in ischaemic myocardium. *Nature 428*:668-673.

[35] Murry, C.E., Soonpaa, M.H., Reinecke, H., Nakajima, H., Nakajima, H.O., Rubart, M., Pasumarthi, K.B., Virag, J.I., Bartelmez, S.H., Poppa, V., et al. 2004. Haematopoietic stem cells do not transdifferentiate into cardiac myocytes in myocardial infarcts. *Nature 428*:664-668.

[36] Chien, K.R. 2004. Stem cells: lost in translation. *Nature 428*:607-608.

[37] Hamano, K., Nishida, M., Hirata, K., Mikamo, A., Li, T.S., Harada, M., Miura, T., Matsuzaki, M., and Esato, K. 2001. Local implantation of autologous bone marrow cells for therapeutic angiogenesis in patients with ischemic heart disease: clinical trial and preliminary results. *Jpn Circ J 65*:845-847.

[38] Fuchs, S., Satler, L.F., Kornowski, R., Okubagzi, P., Weisz, G., Baffour, R., Waksman, R., Weissman, N.J., Cerqueira, M., Leon, M.B., et al. 2003. Catheter-based autologous bone marrow myocardial injection in no-option patients with advanced coronary artery disease: a feasibility study. *J Am Coll Cardiol 41*:1721-1724.

[39] Perin, E.C., Dohmann, H.F., Borojevic, R., Silva, S.A., Sousa, A.L., Mesquita, C.T., Rossi, M.I., Carvalho, A.C., Dutra, H.S., Dohmann, H.J., et al. 2003. Transendocardial, autologous bone marrow cell transplantation for severe, chronic ischemic heart failure. *Circulation 107*:2294-2302.

[40] Stamm, C., Westphal, B., Kleine, H.D., Petzsch, M., Kittner, C., Klinge, H., Schumichen, C., Nienaber, C.A., Freund, M., and Steinhoff, G. 2003. Autologous bone-marrow stem-cell transplantation for myocardial regeneration. *Lancet 361*:45-46.

[41] Goodell, M.A., Brose, K., Paradis, G., Conner, A.S., and Mulligan, R.C. 1996. Isolation and functional properties of murine hematopoietic stem cells that are replicating in vivo. *J Exp Med 183*:1797-1806.

[42] Orlic, D., Fischer, R., Nishikawa, S., Nienhuis, A.W., and Bodine, D.M. 1993. Purification and characterization of heterogeneous pluripotent hematopoietic stem cell populations expressing high levels of c-kit receptor. *Blood 82*:762-770.

[43] Jackson, K.A., Majka, S.M., Wang, H., Pocius, J., Hartley, C.J., Majesky, M.W., Entman, M.L., Michael, L.H., Hirschi, K.K., and Goodell, M.A. 2001. Regeneration of ischemic cardiac muscle and vascular endothelium by adult stem cells. *J Clin Invest 107*:1395-1402.

[44] Gilbert, S.F. 1997. *Developmental Biology*, Fourth Edition. Sunderland, MA: Sinauer Associates, Inc. 342 pp.

[45] Flamme, I., and Risau, W. 1992. Induction of vasculogenesis and hematopoiesis in vitro. *Development 116*:435-439.

[46] Risau, W., and Flamme, I. 1995. Vasculogenesis. *Annual Review of Cell Developmental Biology 11*:73-91.

[47] Carmeliet, P., and Jain, R.K. 2000. Angiogenesis in cancer and other diseases. Nature 407:249-257.

[48] Carmeliet, P. 2003. Angiogenesis in health and disease. *Nat Med 9*:653-660.

[49] Isner, J.M., and Asahara, T. 1999. Angiogenesis and vasculogenesis as therapeutic strategies for postnatal neovascularization. *J Clin Invest 103*:1231-1236.

[50] Asahara, T., Murohara, T., Sullivan, A., Silver, M., van der Zee, R., Li, T., Witzenbichler, B., Schatteman, G., and Isner, J.M. 1997. Isolation of putative progenitor endothelial cells for angiogenesis. *Science 275*:964-967.

[51] Shi, Q., Rafii, S., Wu, M.H., Wijelath, E.S., Yu, C., Ishida, A., Fujita, Y., Kothari, S., Mohle, R., Sauvage, L.R., et al. 1998. Evidence for circulating bone marrow-derived endothelial cells. *Blood 92*:362-367.

[52] Asahara, T., Takahashi, T., Masuda, H., Kalka, C., Chen, D., Iwaguro, H., Inai, Y., Silver, M., and Isner, J.M. 1999. VEGF contributes to postnatal neovascularization by mobilizing bone marrow-derived endothelial progenitor cells. *EMBO J. 18*:3964-3972.

[53] Takahashi, T., Kalka, C., Masuda, H., Chen, D., Silver, M., Kearney, M., Magner, M., Isner, J.M., and Asahara, T. 1999. Ischemia- and cytokine-induced mobilization of bone marrow-derived endothelial progenitor cells for neovascularization. *Nat Med 5*:434-438.

[54] Murohara, T., Ikeda, H., Duan, J., Shintani, S., Sasaki, K.-I., Eguchi, H., Onitsuka, I., Matsui, K., and Imaizumi, T. 2000. Transplanted cord blood-derived endothelial precursor cells augment postnatal neovascularization. *Journal of Clinical Investigation 105*:1527-1536.

[55] Rafii, S. 2000. Circulating endothelial precursors: mystery, reality, and promise. *Journal of Clinical Investigation 105*:17-19.

[56] Kawamoto, A., Gwon, H.C., Iwaguro, H., Yamaguchi, J.I., Uchida, S., Masuda, H., Silver, M., Ma, H., Kearney, M., Isner, J.M., et al. 2001. Therapeutic potential of ex vivo expanded endothelial progenitor cells for myocardial ischemia. *Circulation 103*:634-637.

[57] Kocher, A.A., Schuster, M.D., Szabolcs, M.J., Takuma, S., Burkhoff, D., Wang, J., Homma, S., Edwards, N.M., and Itescu, S. 2001. Neovascularization of ischemic myocardium by human bone-marrow-derived angioblasts prevents cardiomyocyte apoptosis, reduces remodeling and improves cardiac function. *Nat Med 7*:430-436.

[58] Kawamoto, A., Tkebuchava, T., Yamaguchi, J., Nishimura, H., Yoon, Y.S., Milliken, C., Uchida, S., Masuo, O., Iwaguro, H., Ma, H., et al. 2003. Intramyocardial transplantation of autologous endothelial progenitor cells for therapeutic neovascularization of myocardial ischemia. *Circulation 107*:461-468.

[59] Prockop, D.J. 1997. Marrow stromal cells as stem cells for nonhematopoietic tissues. *Science 276*:71-74.

[60] Pittenger, M.F., Mackay, A.M., Beck, S.C., Jaiswal, R.K., Douglas, R., Mosca, J.D., Moorman, M.A., Simonetti, D.W., Craig, S., and Marshak, D.R. 1999. Multilineage potential of adult human mesenchymal stem cells. *Science 284*:143-147.

[61] Makino, S., Fukuda, K., Miyoshi, S., Konishi, F., Kodama, H., Pan, J., Sano, M., Takahashi, T., Hori, S., Abe, H., et al. 1999. Cardiomyocytes can be generated from marrow stromal cells in vitro. *J Clin Invest 103*:697-705.

[62] Toma, C., Pittenger, M.F., Cahill, K.S., Byrne, B.J., and Kessler, P.D. 2002. Human mesenchymal stem cells differentiate to a cardiomyocyte phenotype in the adult murine heart. *Circulation 105*:93-98.

[63] Wang, J.S., Shum-Tim, D., Galipeau, J., Chedrawy, E., Eliopoulos, N., and Chiu, R.C. 2000. Marrow stromal cells for cellular cardiomyoplasty: feasibility and potential clinical advantages. *J Thorac Cardiovasc Surg 120*:999-1005.

[64] Shake, J.G., Gruber, P.J., Baumgartner, W.A., Senechal, G., Meyers, J., Redmond, J.M., Pittenger, M.F., and Martin, B.J. 2002. Mesenchymal stem cell implantation in a

swine myocardial infarct model: engraftment and functional effects. *Ann Thorac Surg 73*:1919-1925; discussion 1926.

[65] Dai, W., Hale, S.L., Martin, B.J., Kuang, J.Q., Dow, J.S., Wold, L.E., and Kloner, R.A. 2005. Allogeneic mesenchymal stem cell transplantation in postinfarcted rat myocardium: short- and long-term effects. *Circulation 112*:214-223.

[66] Tang, Y.L., Tang, Y., Zhang, Y.C., Qian, K., Shen, L., and Phillips, M.I. 2005. Improved graft mesenchymal stem cell survival in ischemic heart with a hypoxia-regulated heme oxygenase-1 vector. *J Am Coll Cardiol 46*:1339-1350.

[67] Mangi, A.A., Noiseux, N., Kong, D., He, H., Rezvani, M., Ingwall, J.S., and Dzau, V.J. 2003. Mesenchymal stem cells modified with Akt prevent remodeling and restore performance of infarcted hearts. *Nat Med 9*:1195-1201.

[68] Gnecchi, M., He, H., Liang, O.D., Melo, L.G., Morello, F., Mu, H., Noiseux, N., Zhang, L., Pratt, R.E., Ingwall, J.S., et al. 2005. Paracrine action accounts for marked protection of ischemic heart by Akt-modified mesenchymal stem cells. *Nat Med 11*:367-368.

[69] Kinnaird, T., Stabile, E., Burnett, M.S., Lee, C.W., Barr, S., Fuchs, S., and Epstein, S.E. 2004. Marrow-derived stromal cells express genes encoding a broad spectrum of arteriogenic cytokines and promote in vitro and in vivo arteriogenesis through paracrine mechanisms. *Circ Res 94*:678-685.

[70] Kinnaird, T., Stabile, E., Burnett, M.S., and Epstein, S.E. 2004. Bone-marrow-derived cells for enhancing collateral development: mechanisms, animal data, and initial clinical experiences. *Circ Res 95*:354-363.

[71] Kinnaird, T., Stabile, E., Burnett, M.S., Shou, M., Lee, C.W., Barr, S., Fuchs, S., and Epstein, S.E. 2004. Local delivery of marrow-derived stromal cells augments collateral perfusion through paracrine mechanisms. *Circulation 109*:1543-1549.

[72] Tang, Y.L., Zhao, Q., Zhang, Y.C., Cheng, L., Liu, M., Shi, J., Yang, Y.Z., Pan, C., Ge, J., and Phillips, M.I. 2004. Autologous mesenchymal stem cell transplantation induce VEGF and neovascularization in ischemic myocardium. *Regul Pept 117*:3-10.

[73] Reyes, M., Lund, T., Lenvik, T., Aguiar, D., Koodie, L., and Verfaillie, C.M. 2001. Purification and ex vivo expansion of postnatal human marrow mesodermal progenitor cells. *Blood 98*:2615-2625.

[74] Reyes, M., Dudek, A., Jahagirdar, B., Koodie, L., Marker, P.H., and Verfaillie, C.M. 2002. Origin of endothelial progenitors in human postnatal bone marrow. *J Clin Invest 109*:337-346.

[75] Jiang, Y., Jahagirdar, B.N., Reinhardt, R.L., Schwartz, R.E., Keene, C.D., Ortiz-Gonzalez, X.R., Reyes, M., Lenvik, T., Lund, T., Blackstad, M., et al. 2002. Pluripotency of mesenchymal stem cells derived from adult marrow. *Nature 418*:41-49.

[76] Schwartz, R.E., Reyes, M., Koodie, L., Jiang, Y., Blackstad, M., Lund, T., Lenvik, T., Johnson, S., Hu, W.S., and Verfaillie, C.M. 2002. Multipotent adult progenitor cells from bone marrow differentiate into functional hepatocyte-like cells. *J Clin Invest 109*:1291-1302.

[77] Yoon, Y.S., Wecker, A., Heyd, L., Park, J.S., Tkebuchava, T., Kusano, K., Hanley, A., Scadova, H., Qin, G., Cha, D.H., et al. 2005. Clonally expanded novel multipotent stem

cells from human bone marrow regenerate myocardium after myocardial infarction. *J Clin Invest 115*:326-338.

[78] Lee, N., Thorne, T., Losordo, D.W., and Yoon, Y.S. 2005. Repair of ischemic heart disease with novel bone marrow-derived multipotent stem cells. *Cell Cycle 4*:861-864.

[79] Wagers, A.J., and Weissman, I.L. 2004. Plasticity of adult stem cells. *Cell 116*:639-648.

[80] Ziegelhoeffer, T., Fernandez, B., Kostin, S., Heil, M., Voswinckel, R., Helisch, A., and Schaper, W. 2004. Bone marrow-derived cells do not incorporate into the adult growing vasculature. *Circ Res 94*:230-238.

[81] Terada, N., Hamazaki, T., Oka, M., Hoki, M., Mastalerz, D.M., Nakano, Y., Meyer, E.M., Morel, L., Petersen, B.E., and Scott, E.W. 2002. Bone marrow cells adopt the phenotype of other cells by spontaneous cell fusion. *Nature 416*:542-545.

[82] Ying, Q.L., Nichols, J., Evans, E.P., and Smith, A.G. 2002. Changing potency by spontaneous fusion. *Nature 416*:545-548.

[83] Orlic, D., Kajstura, J., Chimenti, S., Limana, F., Jakoniuk, I., Quaini, F., Nadal-Ginard, B., Bodine, D.M., Leri, A., and Anversa, P. 2001. Mobilized bone marrow cells repair the infarcted heart, improving function and survival. *Proc Natl Acad Sci U S A 98*:10344-10349.

[84] Nygren, J.M., Jovinge, S., Breitbach, M., Sawen, P., Roll, W., Hescheler, J., Taneera, J., Fleischmann, B.K., and Jacobsen, S.E. 2004. Bone marrow-derived hematopoietic cells generate cardiomyocytes at a low frequency through cell fusion, but not transdifferentiation. *Nat Med 10*:494-501.

[85] Alvarez-Dolado, M., Pardal, R., Garcia-Verdugo, J.M., Fike, J.R., Lee, H.O., Pfeffer, K., Lois, C., Morrison, S.J., and Alvarez-Buylla, A. 2003. Fusion of bone-marrow-derived cells with Purkinje neurons, cardiomyocytes and hepatocytes. *Nature 425*:968-973.

[86] Rehman, J., Li, J., Orschell, C.M., and March, K.L. 2003. Peripheral blood "endothelial progenitor cells" are derived from monocyte/macrophages and secrete angiogenic growth factors. *Circulation 107*:1164-1169.

[87] Hur, J., Yoon, C.H., Kim, H.S., Choi, J.H., Kang, H.J., Hwang, K.K., Oh, B.H., Lee, M.M., and Park, Y.B. 2004. Characterization of two types of endothelial progenitor cells and their different contributions to neovasculogenesis. *Arterioscler Thromb Vasc Biol 24*:288-293.

[88] Urbich, C., Aicher, A., Heeschen, C., Dernbach, E., Hofmann, W.K., Zeiher, A.M., and Dimmeler, S. 2005. Soluble factors released by endothelial progenitor cells promote migration of endothelial cells and cardiac resident progenitor cells. *J Mol Cell Cardiol 39*:733-742.

INDEX

A

B

C

D

E

F

I

N

O

P

Q

R

T

U

V

W

X

Y

Z